Phytochrome and Photomorphogenesis

The effect of white light on the pattern of development of bean seedlings. The seedling on the left has been grown for six days in total darkness; the seedling on the right is genetically almost identical, but has been grown for six days in white light.

Phytochrome and Photomorphogenesis

An Introduction to the Photocontrol of Plant Development

Harry Smith

Professor of Plant Physiology
University of Nottingham School of Agriculture

London · New York · St Louis · San Francisco · Düsseldorf · Johannesburg
Kuala Lumpur · Mexico · Montreal · New Delhi · Panama · Paris · São Paulo
Singapore · Sydney · Toronto

Published by McGRAW-HILL Book Company (UK) Limited

MAIDENHEAD · BERKSHIRE · ENGLAND

Library of Congress Cataloging in Publication Data

Smith, Harry, Date
   Phytochrome and photomorphogenesis.

   Bibliography: p. 223
   1. Plants, Effect of light on. 2. Phytochrome. 3. Plant morphogenesis. I. Title.
[DNLM: 1. Photochemistry. 2. Plants — Growth and development.
QK899 S649p 1974]
QK757.S64   581.1'9'153   74–5349
ISBN 0–07–084038–5

PRINTED AND BOUND IN GREAT BRITAIN

for Elinor

Contents

Preface

Light is the ultimate source of free energy for virtually all life on our planet. Solar energy is captured by the chlorophyll of the green plant and transformed, through photosynthesis, into potential chemical energy in the form of complex organic substances which, directly or indirectly, form the food sources for heterotrophic organisms. Through this process, the green plant constitutes a fundamental, inescapable link between the biosphere and its energy supplies. The importance of light was apparent even to early mankind as is evident from the ancient religions of sun-worship, in which the superstitions show a primitive realization of a life-giving force. In the present, hopefully less superstitious, times, the role of the green plant as the transducer of this vital energy is fully recognized, although the molecular mechanisms involved may still appear mystical to many. This book is not, however, concerned with the capture of solar energy. Light has many other important functions in the life of the green plant, apart from the provision of energy. In particular, it is one of several environmental factors which regulate plant growth and development.

The processes of development in plants differ in principle from those in animals in that they are highly plastic. In other words, morphogenesis in plants is susceptible to quantitative and qualitative modification by external influences, a fact which presents the biologist with an unparalleled opportunity to delve into the molecular mechanisms through which development is regulated. Treatment with light is a particularly useful research tool since, in contrast to chemical and hormonal treatments, for example, light can be applied and removed without any fear of complications resulting from residual amounts of the exogenous stimulus being left in the tissue. As will be seen, examples exist of the light regulation of almost every phase of plant development, and the study of these regulatory processes has already contributed a great deal to our understanding of the mechanisms of developmental regulation in plants. The elucidation of the molecular basis of differentiation and development is the next major problem confronting biologists, and it can safely be predicted that the study of the photocontrol of plant development will have an increasingly important role in this extension of knowledge.

Photobiology is currently an extremely exciting field of biological research with far-reaching discoveries being made regularly in the fields of vision, photosynthesis, and photomorphogenesis. In all these cases, success can only be achieved through the day to day partnership of physicists, physical chemists, spectroscopists, biochemists, and biologists. Photobiology is therefore a very useful subject from the educational point of view since an intensive coverage stimulates the student to draw on and to reinforce his general background in physics, chemistry, and biology. It is the author's view that only through such an integrated approach can the subject of photobiology be taught successfully, and this book has been compiled with that principle in mind. It probably falls far short of the intended ideal, since the author is a mere biologist and not an expert in the physical and chemical aspects of the subject. Nevertheless, if it stimulates a few physical scientists and a few biologists to develop an interest in each other's subject, this book will have achieved its purpose.

Photomorphogenesis is treated here at an introductory level, specifically for final year undergraduates and postgraduate students starting study of botany, biology, biochemistry, horticulture, and agriculture, and for students of other disciplines, e.g., physics and physical chemistry, who may have aspirations towards research in photobiology. It should also be of use to college and university teachers attempting to get to grips with the vast amount of information available on the subject. In chapter 2 the biologist is introduced to the fundamental nature of light as it affects living organisms. In chapters 3 to 5, the known photoreactive systems are analysed, as far as possible at the molecular level. The final four chapters are an attempted synthesis to show how the reaction systems operate to control the growth and development of the plant. No attempt has been made to be encyclopaedic in the coverage of this question; rather, selected aspects of development are dealt with in depth to illustrate the mainstream of current thought and speculation. The molecular approach has been used throughout the whole book, since the author feels that only by pinpointing mechanisms at the biochemical level can real precision and understanding be achieved. Already, the study of photomorphogenesis is progressing to the biophysical level, but full understanding of the overall process will only be attainable when the physiology and biochemistry of the photoreactive systems are adequately described. This book is an attempt to interest aspiring research workers in a subject of great fundamental and practical importance, and one which transcends the boundaries of biology, biochemistry, and biophysics.

I am grateful to many colleagues for helpful discussions during the writing of this book; in particular I wish to thank Professor Malcolm Wilkins, Consulting Editor for the series, and Professor P. F. Wareing, FRS, for the advice they have freely given from time to time. I am also grateful to Professor J. L. Monteith, FRS, for discussions concerning the content and accuracy of chapter 2, and Dr R. E. Kendrick for helpful advice and for much of the information contained in the Appendix. The major part of the typescript was prepared by Miss Joyce Shore and Mrs Olive Davison of the Botany Department at Manchester, and I am grateful for their patience while dealing with me at a distance. Thanks are also due to Miss M. J. Bentley and Mrs Susan Smith who typed some sections at Sutton Bonington. This book has taken me a long time to produce from the first thought to the final typescript; in all this time I have been supported by my wife Elinor without whose fortitude it would never have been finished.

H. Smith
1 *June* 1973

Acknowledgements

The following publishers have generously given permission to publish the stated figures: Academic Press Inc., Fig. 3.1; American Chemical Society, Figs 4.10a, 4.14; American Journal of Botany, Figs 7.1, 8.2, 8.3; American Society of Plant Physiologists, Figs 3.16, 4.6, 4.9, 4.15, 5.1, 5.10, 5.12, 5.14, 5.16, 6.1b, 7.9, 8.1, 8.26, 8.27; Cambridge University Press, Fig. 3.7; Centrex Publishing Co., Fig. 2.6; Elsevier Pub. Co., Fig. 8.15; Macmillan Journals Ltd., Figs 5.2, 5.7; McGraw-Hill Bk. Co., Ltd., Figs 3.11, 3.13, 3.14, 5.4b, 5.5b, 5.6, 5.11, 5.17; Pergamon Press, Figs 3.3, 3.5, 3.6, 3.10, 4.10b, 4.11a, 4.11b, 5.9, 8.4, 8.14; Scandinavian Society for Plant Physiology, Fig. 5.15; Springer-Verlag, Figs 5.13c, 5.18, 6.3, 6.5, 6.6, 6.7, 7.7, 7.8, 8.12, 8.17, 8.18, 8.19; Verlag der Zeitschrift für Naturforschung, Figs 3.9, 8.7; Weizmann Science Press, Fig. 6.4.

Beans and many other plants, which stand where they are much shaded, being thereby kept continually moist, do grow to unusual heights, and are drawn up as they call it, by the overshadowing trees, their parts being kept long, soft and ductile.

Stephen Hales (*Statical Essays*, **1**, 334, 1727).

1
Introduction – definitions and scope

In a natural environment, all organisms compete in the struggle for survival. In the case of the young seedling, developing in the midst of others which may already have an advantage over it, the struggle is essentially for sunlight. Solar irradiation provides the developing seedling with the energy it requires to manufacture new cellular materials, and also regulates and modulates the development of the plant throughout its life. Virtually every phase of plant development, and many other phenomena not intimately involved in development, is subject to photocontrol independently of photosynthesis. This distinction is very important; photosynthesis provides the energy for plant growth and development but does not in any major way regulate the pattern of development. Completely separate photoresponsive mechanisms have been evolved through which plants adapt to the radiation environment. It is with these purely developmental responses that we are concerned in this book.

1.1 Definitions

In common with other rapidly advancing disciplines, progress in the study of the photocontrol of plant development has outstripped the semantic inventiveness of the research workers in the field, and the literature is consequently laden with ill-defined, ambiguous, and overlapping 'photophenomena'. Before we can proceed further it is necessary to cut a clearing in this terminological jungle and to define a limited set of light-regulated developmental processes.

We can first restrict our consideration to those processes which can be strictly defined as being developmental in nature. This criterion excludes such non-developmental processes as photosynthesis, photorespiration, photonasty, and stomatal movements. It should be noted at the outset, however, that some of these non-developmental phenomena appear to be regulated by precisely the same molecular mechanisms as are important in the photocontrol of development, and therefore from time to time reference will be made to them, where such consideration is beneficial to the main theme. It should also be remembered that many metabolic processes, such as the mobilization of storage reserves in the germinating seed, may not, at first sight, be recognized as developmental processes, yet nevertheless they are absolutely essential to the development of the organism and may, themselves, be regulated by light.

We are concerned, therefore, only with the photocontrol of development. Even here, however, some restriction of scope is necessary if this book is not to become impossibly long and tedious. Fortunately, light exerts its control over development in several different ways, allowing the restriction of scope to be more or less logical.

The criteria we use in defining the different ways in which light regulates development involve a consideration of the light stimulus required and of the characteristics of the response evoked. These criteria are:

1. whether the light stimulus must be directional;
2. whether the light stimulus must be periodic; and
3. whether the developmental response is directional.

Using these criteria we can recognize three major categories of light responses:

1. *Phototropism* – a directional growth response to a directional light stimulus.
2. *Photoperiodism* – a non-directional developmental response to a non-directional, periodic light stimulus.
3. *Photomorphogenesis* – a non-directional developmental response to a non-directional, non-periodic light stimulus.

The decision-making process used to arrive at these definitions is shown schematically in Fig. 1.1.

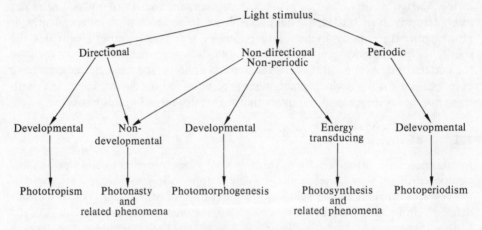

Fig. 1.1 How to define plant responses to light.

Within each of these categories, various sub-categories can be defined, based on energy requirements, wavelength regions of maximum activity, timing requirements, etc. Such fine analysis, however, is best left to later chapters.

There are therefore three major types of developmental photoresponses: phototropism, photoperiodism, and photomorphogenesis. Each of these has its own vitally important part to play in the developmental adaptation to the environment.

(a) Phototropism

Phototropism is here defined as a directional growth response to a directional light stimulus. This means that a plant, or a part of a plant, which is irradiated unilaterally with light of actinic (i.e., photochemically active) wavelengths, will respond by growth in a direction related to the direction of the incident irradiation. Although strictly outside the scope of this book, phototropism is treated in some detail in chapter 3 since many of its more important features reflect aspects of photomorphogenesis.

In the natural environment, phototropism is probably of great importance in orienting the growth of the plant, so that it is able to optimize its interception of the available solar energy for photosynthesis. Such a mechanism would obviously provide a plant with a great adaptive advantage over non-phototropic plants, and it is known from comparative studies that phototropism probably appeared very early in the evolution of land plants.

The orientation of leaves with respect to sunlight on a day to day basis is not brought about by phototropism, since the curvatures due to phototropic growth changes are fixed and irreversible. Leaf movements, together with those of petals and bracts, are mainly due to changes in turgor pressure in certain specialized pulvinal or 'hinge' cells situated at the base of the petiole or midrib. Such movements are known as nastic movements, and when mediated by light the phenomenon is termed photonasty. Again, although photonasty is strictly outside the scope of this book, it is mentioned in passing at various points as it has important implications for our understanding of photomorphogenesis.

(b) Photoperiodism

Photoperiodism is defined here as a non-directional developmental response to a non-directional, periodic light stimulus. This means that when a plant is subjected to a repeated sequence of periodic light stimuli with specific time characteristics, it responds by specific developmental changes. The most intensively investigated photoperiodic response is flowering, through the study of which photoperiodism was first recognized. Several other developmental processes also come under photoperiodic control in certain species including seed germination, bud dormancy, and the rate and pattern of vegetative development.

Photoperiodism is the mechanism which enables plants to adapt to the seasonal nature of the environment in the temperate and sub-arctic regions. Thus, many woody plants initiate dormant winter buds during late summer when the climatic conditions are still favourable for continued growth. The lengthening of the daily dark periods in these species evokes a developmental response which anticipates the harsh conditions of winter. In the case of flowering, this adaptive mechanism has been refined in many different ways so that certain species will only flower when the daily dark periods are longer than a certain critical period (these plants are known as short-day plants); others will only flower when the daily dark period is shorter than a certain critical period (long-day plants); while some others require a sequence of both types of stimuli before they will flower (long-day–short-day plants). In each case, the particular response can be seen to act as an adaptive mechanism for determining optimum time for flowering and subsequent seed set.

Photoperiodism has been studied in great depth since it was first discovered by Garner and Allard[151] in 1920. It is, however, strictly beyond the scope of this book as defined by the title and in any case is admirably covered in a companion volume published in the same series (*Photoperiodism in Plants*, by Dr Daphne Prue[343]).

Photomorphogenesis and photoperiodism do overlap significantly, but it seems true to say that consideration of photoperiodism does not, at present, yield any information which is particularly useful for an understanding of photomorphogenesis. Thus, the omission of photoperiodism from the coverage here is not seriously detrimental to the understanding of photomorphogenesis.

(c) Photomorphogenesis

Photomorphogenesis is defined here as a non-directional developmental response to a non-directional, non-periodic light stimulus. This makes photomorphogenesis a somewhat garbage can subject, since it includes all those ways in which light regulates development which are not covered by the more precisely defined phenomena of phototropism and photoperiodism. Moreover, the above definition would probably not receive universal approval since certain workers have tended to use a much broader definition including photoperiodism and phototropism within its frame of reference. Professor Hans Mohr, of the University of Freiburg, for example, has defined photomorphogenesis as 'the control which may be exerted by light over the growth, development, and differentiation of a plant, independently of photosynthesis.[293] It is the opinion of this author that such a definition is too broad to be particularly helpful, especially when we are attempting to analyse the molecular mechanisms underlying the processes.

From the adaptational point of view, it is obviously of crucial importance for the germinating seed, or the developing seedling, to have a mechanism which enables it to react to the presence or absence of light in an appropriate manner. Speaking anthropomorphically, the young plant would then 'know' whether it was situated beneath the soil, or on the surface, and be able to develop accordingly. It also seems likely that adaptive advantage would be conferred on any plant which had a mechanism for detecting shading from another plant, especially if this mechanism were able to bring about a changed pattern of growth such that the shaded plant grew away from its neighbour.

As we shall see, photomorphogenic responses occur throughout the whole life of the plant and may be responsible for controlling seed germination, stem elongation, leaf expansion, phototropic and geotropic sensitivity, the formation of stomata and leaf hairs, the synthesis of chlorophyll, the development of chloroplast structure, and the synthesis of a wide range of secondary products including pigments such as the anthocyanins.

Clearly, not all of these responses are regulated by light at the same time in any one species, and also quite clearly, several possible responses are omitted from this list. The morphogenic importance of light can perhaps best be appreciated by inspection of the Frontispiece, which shows a pair of genetically identical bean plants grown for the same length of time under identical conditions – with the exception that the plant on the left was grown in complete darkness, while that on the right was grown in daylight. The striking differences between the two seedlings will not be described in detail here since they form an important part of chapter 7; a moment or two's contemplation, however, will give the reader a feeling for the undeniable importance of light in the life of plants.

1.2 The fundamental nature of photomorphogenic reactions

In any biological response evoked by light energy a basic sequence of events must occur. The light must be absorbed by a specific photoreceptive molecule whose chemical reactivity is thereby changed. By virtue of its changed chemical reactivity the photoreceptor then initiates a sequence of metabolic processes which lead ultimately to the observed developmental changes (see Fig. 1.2). In order fully to understand photomorphogenesis we therefore need to acquire information on:

1. the nature of the required light stimulus;
2. the chemical identity of the photoreceptor and the photochemical changes which occur upon light absorption;
3. the immediate mechanism through which the photoreceptor triggers the developmental responses;
4. the nature of the ultimate developmental changes; and
5. the molecular and biochemical nature of the partial processes that intervene between the photoreceptor and the ultimate morphogenic phenomena.

An attempt has been made in this book to cover all these processes, more or less as they occur in the plant. The subject is enormous, however, with at least 2000 papers published on photomorphogenesis alone in the last decade, and therefore the coverage will be less than complete so that the important principles do not become submerged in a sea of unrelated facts.

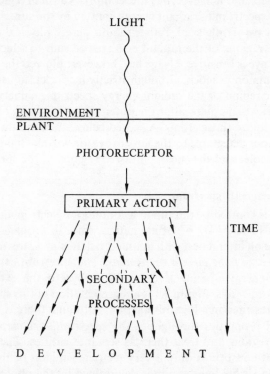

Fig. 1.2 Partial processes in a photomorphogenic response.

2
Light as a biological agent

2.1 The physical nature of light

Light, as we understand it in everyday life, comprises a small region of the continuous electromagnetic spectrum of radiant energy emitted by the sun. Strictly speaking, the term 'light' is a psychophysical, rather than a purely physical term, and it can be defined as those regions of the radiant energy spectrum to which the average light-adapted human eye is sensitive. Usage has, however, blurred the edges of this definition and it is common practice, although strictly incorrect, to use the term 'light' to describe a wider region of the radiant energy spectrum, namely those wavelengths possessing sufficient energy to alter the outer electronic energy levels of atoms, but not sufficient to ionize those atoms. As noted below, the term therefore encompasses not only the regions detectable by the human eye, or 'visible' light, but it also includes the 'near' ultraviolet and the 'near' infra-red regions of the spectrum.

(a) The electromagnetic spectrum

Radiant energy is emitted by the sun in a continuous electromagnetic spectrum. The properties of this energy are of a dual nature, since in propagation it behaves as a waveform, while on interaction with matter it behaves as a stream of discrete packets of energy known as *light quanta* or *photons*. The spectrum extends from the very long-wave, low-energy quanta of the radio waves to the extremely high-energy quanta of the cosmic rays. Figure 2.1 is a representation of this spectrum in which the wavelength is presented in a logarithmic plot. Although there is a continuous transition in physical properties throughout the spectrum, it is common to arbitrarily delimit certain regions and treat them as separate entities. The diagram illustrates how visible light constitutes only a very restricted region of the overall spectrum. The range of the visible light region is determined by the photochemical properties of the human visual pigments, and extends from about 380 nm to about 770 nm wavelength.

As light is propagated through space as a wave, it is possible to derive three characteristic properties of any particular wave:

1. the *wavelength*, or the absolute distance between two consecutive crests in the waveform;
2. the *frequency*, or the number of wave crests passing a stationary point in one second; and
3. the *wavenumber*, or the number of wave crests present in a one centimetre length of the wave.

These three parameters are related by the equation:

$$\sigma = 1/\lambda = v/c$$

where:

σ = wavenumber
v = frequency
λ = the wavelength and
c = velocity of light

Wavelength is measured in ångström units (Å or 10^{-10} m) or more correctly in nanometers (nm or 10^{-9} m), formerly known as millimicrons (mμ). Frequency is measured in cycles per second, while wavenumber is expressed in reciprocal centimetre (cm^{-1}).

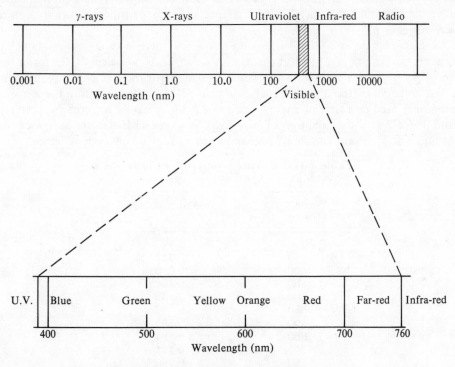

Fig. 2.1 The electromagnetic spectrum of solar radiation.

(b) The quantum nature of light

When light is absorbed by matter it behaves as a stream of discrete indivisible particles, or packets, of energy, in contrast to the wavelike properties characteristic of the propagation of light. This duality is found throughout the whole sphere of particle physics. Electrons, protons, neutrons, and photons have certain properties best described in wave terms, and others more easily understood in terms of a stream of particles. Thus, the propagation of light is best considered in wave terms, while its absorption is best thought of in particle terms.

Radiant energy may therefore be characterized by wavelength, frequency, or wavenumber, and also by the energy of the quantum, or photon, its smallest component.

The quantum theory states that the energy of the quantum is directly proportional to the frequency, the constant of proportionality being Planck's constant, h, which has a numerical value of $6 \cdot 624 \times 10^{-7}$ erg s. The energy of a quantum is therefore related to the frequency of the corresponding waveform by the equation:

$$E = h\nu = h\sigma c = hc/\lambda$$

The energy of a single quantum or photon is a very small number and for practical purposes it is simpler to use a larger multiple. The multiple chosen is the Einstein, E, which is the energy of a mole of quanta. Expressed algebraically:

$$E = Nh\nu$$

where N is Avogadro's number, $6 \cdot 02 \times 10^{23}$ molecules per gram molecule. Although radiant energy can be expressed in absolute energy units, the preferred SI unit being the joule, the use of the Einstein is becoming more common in biological literature. Since the Einstein has a very large value, quantum flux densities are often given in picoEinsteins ($pE = 10^{-12}$ E).

As the energy per quantum varies with frequency, there is an inverse relationship between wavelength and quantum energy – the longer the wavelength, the lower the energy per photon. This is shown in Table 2.1 which lists the energy per quantum in terms of the *joule* (the absolute energy unit in SI). In older literature the erg was commonly used as a unit of energy, one joule being equal to 10^7 ergs. We can see from Table 2.1 that a quantum of blue light of wavelength 400 nm has twice the absolute energy of a quantum of near infra-red light of wavelength 800 nm.

Blue light more energy.

Table 2.1 Quantum energy of various wavelengths of radiant energy.

Wavelength nm	Quantum energy J/µE*
300	0·399
350	0·342
400	0·299
450	0·266
500	0·239
550	0·217
600	0·199
650	0·184
700	0·171
750	0·160
800	0·150

* 1 Joule $\equiv 10^7$ ergs.
1 Einstein $\equiv 6 \cdot 02 \times 10^{23}$ quanta.
1 µE $\equiv 10^{-6}$ Einsteins.

The importance of expressing radiant energy in quantum units comes from the application of Einstein's law of photochemical equivalence which states that in the primary photochemical reaction one quantum must be absorbed for every atom or molecule reacting. Thus, for the photochemical activation of one gram molecule of an absorbing substance, one Einstein of light is required. In experiments involving the comparison of the effects of different wavelengths on biological systems, it is therefore advisable to express absorbed energies in Einsteins, since only then can direct comparisons of response magnitudes be made.

(c) Terminology

Some of the terms used in photobiology together with their definitions are given in Table 2.2. A major source of confusion in the biological literature lies in the incorrect use of the word *intensity* to mean either *irradiance* or *emittance*, depending on the circumstances. Intensity is correctly defined as the flux radiated per unit solid angle, but it is normally used as a term for the flux intercepted per unit area. In this book the correct term, irradiance, is used in all those cases where most biologists loosely use intensity. For reasons stated above concerning the law of photochemical equivalence, it is highly recommended that all irradiances be expressed in quantum terms, and thus quantum flux density should be used in preference to expressions of irradiance in absolute energy units.

Table 2.2 Terms and units commonly used in photobiology (after Withrow and Withrow[466])

Term	SI unit	Definition
Radiant energy	joule	Amount of energy
Radiant flux	watt	Time rate of flow of energy, power
Radiant emittance	watt m^{-2}	Flux radiated per unit area of source
Radiant intensity	watt ω^{-1}*	Flux radiated per unit solid angle
Radiance	watt m^{-2} ω^{-1}	Flux radiated per unit area and solid angle
Irradiance	watt m^{-2}	Flux intercepted per unit area
Quantum flux density	Einsteins sec^{-1} cm^{-2}	Flux intercepted per unit area expressed in quantum units

* ω = solid angle (steradian).

2.2 The absorption of light

As light penetrates an absorbing substance it is attenuated to a degree determined by the probability that individual quanta will be absorbed by atoms or molecules. On the absorption of a photon, the atom or molecule gains all the energy of the photon, and is thereby energized. The consequences of photon absorption depend on the frequency of the radiant energy absorbed. In the far infra-red, the frequency is very low and the energy per quantum is similarly low. Absorption of this radiation can lead only to an increase in the rotational or vibrational energy of the capturing molecules. Alternatively, in the very short-wave gamma rays, X-rays, and far ultra-violet rays the energy per quantum is relatively high and absorption of this energy can result in the complete ejection of an electron from the molecule, thus causing ionization. Between these two extremes is a region, extending from the midultraviolet (*ca*. 290 nm) through the visible, to the near infra-red (*ca*. 800 nm), in which absorption of quanta leads to a change in the energy levels of outer electrons and thus to a photochemical reaction.

Photons are indivisible packets of energy and an electron will absorb a photon only if it is capable of absorbing *all* the energy of that photon. The amount of energy that a particular electron can absorb is determined by the energy levels available to that electron within the molecule. Bohr's theory of atomic and molecular structure states that an atom or molecule can exist only in one of a series of discrete energy levels, and a quantum can be absorbed only if it has just the right amount of energy to raise the atom or molecule from one energy level to a higher, allowable energy level.

Since quantum energy is inversely proportional to wavelength, this relationship implies that an absorbing atom or molecule can absorb light of a specific wavelength only, and thus absorbing substances can be characterized by the wavelengths at which they absorb light. The commonest way of characterizing the absorption of a substance is to measure its *absorption spectrum* – i.e., to construct a plot of the amount of light absorbed against wavelength. This is normally carried out by the use of a spectrophotometer, an instrument which scans a solution of the absorbing substance with a spectrum of monochromatic light beams over a particular wavelength range, and determines the amount of light transmitted at each wavelength.

The amount of light transmitted by a sample of an absorbing substance is described by Beer's law:

$$I = I_0 \, e^{-\mu cb}$$

which can also be written as:

$$\log_e I_0/I = \mu cb$$

where:

I_0 = irradiance of incident beam
I = irradiance of transmitted beam
μ = coefficient of absorption
c = concentration
b = path length.

When the concentration is 1 molar, the path length is 1 cm, and logarithms to the base ten are used in place of natural logarithms, the coefficient of absorption (μ) becomes the molar coefficient of absorption, or the molar extinction coefficient, Σ:

$$\log_{10} I_0/I = \Sigma$$

The ratio I/I_0 is commonly known as *transmittance*, while $\log_{10} I_0/I$ is known as *absorbance*, formerly known as *optical density*. Thus, the absorbance of a solution is directly related to the concentration of that solution.

2.3   The principles of photochemical reactions

All biological processes initiated by light depend on the absorption of light by specific photoreceptive substances which become activated, or energized in the process. Thus, the primary process in a photomorphogenic reaction sequence is a photochemical reaction; that is, a reaction which occurs initially from an electronically excited state of the photoreceptive molecule. For the purposes of this discussion, photochemical reactions can be divided into two sequential processes:

1. the absorption of a photon by an electron and the consequent displacement of the electron to a higher energy level; and
2. the subsequent thermochemical reaction in which the photochemically activated molecule takes part.

The photophysical changes which precede the chemical changes are conventionally described in terms of the Jablonski diagram (Fig. 2.2) in which the possible activated and ground states are depicted by the heavy horizontal lines and energy is represented by the vertical axis. The *ground state*, S_0 is the most stable configuration of the

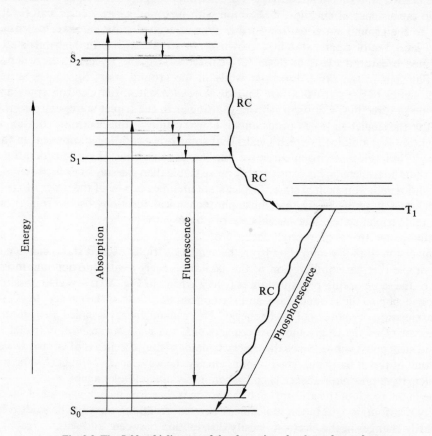

Fig. 2.2 The Jablonski diagram of the absorption of a photon by an electron.

S_0 = ground state (singlet)
S_1, S_2 = singlet excited states
T_1 = triplet excited state
RC (wavy arrows) = radiationless conversions (from Porter.[333] Redrawn from Carl P. Swanson, AN INTRODUCTION TO PHOTOBIOLOGY. © 1969. By permission of Prentice-Hall, Inc., Englewood Cliffs, New Jersey, USA.)

molecule, and absorption of photons of energy e_1 or e_2, for example, energize the molecule to the S_1 or S_2 energy levels, or *excited states*. Superimposed on these major energy levels, molecules of two or more atoms have a series of other, similarly quantized vibrational and rotational states, depicted by the light horizontal lines. Thus the ground state S_0, and the excited states S_1 and S_2 each have a number of other energy levels available to them, and the energy differences between S_0 and the available excited state energy levels determine the size of the quanta that can be absorbed. In its turn, this determines the band width of wavelengths that can be absorbed – in other words, the absorption spectrum of the substance. In complex molecules, i.e., those containing more than ten atoms, the vibrational and rotational states are so numerous and closely spaced that the sharp absorption bands characteristic of the elements do not exist, and the absorption is spread over a broad band of wavelengths. This occurs even with simple molecules in the infra-red region.

The excited states have a relatively short lifetime, roughly 10^{-9} to 10^{-10} seconds and, in the absence of chemical interaction with surrounding molecules, can revert either to the ground state, or to a further type of excited state, of lower potential energy level, but of greater stability, known as a *triplet* state (T_1). In the triplet state, so named because it exists in three very closely spaced energy levels, the valence electrons spin in the same direction, while in the ground state, S_0, or the other excited states (S_1, etc.), which are known as *singlet* states, the electron spins are opposite in direction. Conversion from the singlet to the triplet state, or from the triplet to the singlet state can occur only if conservation of the electronic rotational momentum is maintained, which means that some other rotating component in the molecule, or its close environment, must 'flip' at the same time. These considerations show that the triplet state is considerably more stable than the singlet excited states – yet it still remains an excited state. Triplet state lifetimes can be of the order of 10^{-3} seconds and it seems likely that most photochemical reactions occur from these metastable triplet states. The special case of photosynthesis, in which an electron is actually ejected from chlorophyll occurs from a singlet state.

Conversion to the ground state from the singlet or triplet excited states can occur by a simple thermal equilibration of the excitation energy with surrounding molecules of the same or different substances (wavy arrows in Fig. 2.2). It is also possible for conversion to the ground state, and the concomitant loss of the energy, to occur by the emission of photons of lower energy. This radiant energy is called *fluorescence* in the case of singlet to ground state conversions, and *phosphorescence*, in triplet to ground state conversions. Since the energy content of fluorescence is obviously lower than that of the initially absorbed radiant energy the wavelength is longer, while the wavelength of phosphorescence is correspondingly much longer again.

The initial result of the absorption of a photon is therefore a change in the electron density distribution within the molecule. Chemical reactions can now be said to be primarily changes in the electron density distribution between, and within, reacting molecules, and thus the excited state must be considered as a new chemical species, and not just an energized molecule. The net result of the photophysical processes therefore, is to produce a molecule with a changed chemical reactivity. This fact is of fundamental importance in the study of photomorphogenesis, since it means that the photoreceptive substance can be present under conditions conducive to reaction, but it will not react until light quanta of the correct energy are absorbed.

It is implicit in the above discussion that the overall extent of a photochemical reaction is directly proportional to the total number of quanta absorbed. For any given light treatment, the number of quanta is equal to the quantum flux density (see page 9) multiplied by the time of irradiation in seconds. Thus, in a strictly photochemical reaction, the quantum flux density required to produce a given reaction is inversely proportional to the time. To put it another way, the amount of reaction is proportional to the product of quantum flux density and time. This relationship is known as the Bunsen–Roscoe reciprocity law.

Several photomorphogenic phenomena exhibit reciprocity, and thus demonstrate that the photochemical reaction is the only limiting step. On the other hand, certain other important photomorphogenic processes do not obey the reciprocity law, in that the time of irradiation is more important than quantum flux density in determining the overall extent of the biological responses. In these cases, it seems clear that other rate-limiting steps exist in addition to the photochemical reaction.

2.4 Photochemical reactions in biology

Photochemical reactions are widespread throughout the biosphere; in fact virtually all life on this planet is ultimately dependent on a single photochemical reaction unique to green plants – photosynthesis. The photochemical reactions of photosynthesis also constitute a special case from a biochemical viewpoint in that upon the absorption of light quanta, an electron is apparently ejected from the photoreceptive molecule, chlorophyll, and used in subsequent energy-dependent thermochemical reductive reactions. The chlorophyll molecule is subsequently returned to the ground state by accepting another electron from its immediate environment. Other photochemical reactions in biology rely only on the chemical properties of the excited states of the absorbing molecules, and do not involve a direct utilization of the energy captured. The functional significance of this distinction is that photosynthesis is a mechanism for the capture of radiant energy and its efficient conversion into chemical energy to be used for biochemical syntheses, while the other major photochemical reactions, such as vision and the initial reactions in photomorphogenesis, are not energy gathering reactions *per se*.

(a) Action spectra

The first problem in an understanding of a biological photoreaction is the identification of the responsible photoreceptor substance – often known simply as a *pigment* in the biological literature. From the considerations outlined in the previous section, it is clear that pigment molecules will have characteristic absorption spectra. Furthermore, if the biological photoreaction is a simple function of the excitation of the pigment by light quanta, then the wavelengths of maximum biological activity should correspond to the wavelengths of maximum pigment absorption. In practice, the experimenter constructs a plot of the relative effectiveness of light energy against wavelength (known as an *action spectrum*), and compares this with the absorption spectra of putative photoreceptor substances. In this way, extremely valuable evidence on the identity of the active pigment may be obtained, and this technique has been used with great effect in the study of photomorphogenesis. The correct method for constructing an action spectrum has not, however, been rigorously followed in all cases. It is important to demonstrate that for all wavelengths used the relationship between radiant energy and biological response is similar and preferably linear, so that the energy required to produce a given response can be plotted against wavelength of irradiation.

A less satisfactory method, sometimes used, is to determine the biological response (e.g., percentage germination of batches of seeds) when irradiated with equivalent energies from a range of monochromatic light sources, and thus to plot response against wavelength. This latter method will obviously yield similar results *if* the dose-response relationships are similar throughout the wavelength range, but it does not include a means of assurance on this point, and therefore can give misleading results at times. Cases do arise, of course, when it is, in practice, impossible to perform complete dose-response experiments throughout the wavelength range, and in those instances we must be resigned to the less reliable, but still useful, information derived from the second method.

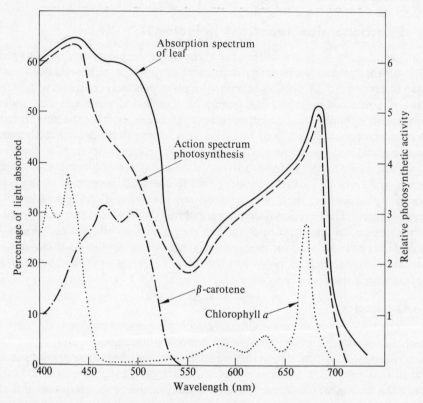

Fig. 2.3 Comparison of the action spectrum for photosynthesis with the absorption spectra of leaves and their pigments (from Kirk and Tilney-Bassett).[239]

Other problems also attend the identification of the active photoreceptor by means of the action spectrum. For example, the pigment may reside deep inside a tissue mass and the radiant energy incident upon it may be attenuated in specific wavelength regions by the absorption of other substances in outer cells or tissues. In this case, the apparent biological response as a function of irradiance would be spuriously small in those wavelength regions attenuated by the absorption in the outer tissues. A further problem is found in the specific case of photosynthesis where pigments other than the photochemical substance, the so-called accessory pigments, can absorb light energy and transfer it by resonance to chlorophyll, the site of the photochemical reaction. In this case, the overall action spectrum approximates to a summation of the absorption spectra of all the pigments involved (Fig. 2.3).

(b) Light and dark reactions in photosynthesis

It is often necessary when investigating the partial processes of a biological photo-response, to determine which reactions are photochemical, i.e., require light, and which are thermochemical, i.e., do not require light. The thermochemical reaction is often spoken of as a 'dark' reaction, but this does not imply that the reaction will not proceed in the light, only that light is not necessary. An important diagnostic feature of a photochemical reaction is that it is not sensitive to temperature as is a thermochemical reaction. Thus, the light reaction in a biochemical sequence will often proceed normally at very low temperatures, at which the dark reactions would occur only very slowly. Photosynthesis presents an excellent example of light and dark reactions acting sequentially. The synthesis of complex, highly reduced organic molecules, such as sucrose, from the more simple, less reduced molecules of carbon dioxide and water requires light in the intact green plant cell. Light energy, however, is only directly involved in the very earliest reactions of this sequence, in which it is used to drive the reduction of nicotinamide adenine dinucleotide phosphate (NADP) to NADPH, and the synthesis of adenosine triphosphate (ATP) from adenosine di-phosphate (ADP) and inorganic phosphate. These two cofactors then drive the reduction of carbon dioxide to the level of carbohydrate by reactions which will proceed in darkness. The synthesis of sucrose from carbon dioxide and water, therefore, is a 'dark' reaction while the reduction of $NADP^+$ ion and the phos-phorylation of ADP are light reactions. In fact, when narrowed down even further, the light reaction in photosynthesis is found to consist of the reduction of one, or possibly two, primary electron acceptors by electrons emitted from chlorophyll molecules under the influence of photon absorption. The reduced primary acceptors then reduce $NADP^+$ and provide the energy for the phosphorylation of ADP by other dark reactions. The actual mechanism is much more complex than this brief description suggests, and almost certainly involves the cooperative action of at least two different photochemical reactions and the removal of electrons from water (see references 154 and 347 for further reading).

(c) Vision

It will also be of use to briefly consider the other major photochemical reaction in biology, vision, since some of its molecular features are of particular significance for the study of photomorphogenesis. Light is received in the human eye by the visual pigments which are found in the rods and cones of the retina. The visual pig-ment known as rhodopsin, or visual purple, consists of a non-absorbing protein moiety, known as opsin, and a light-absorbing non-protein group (known generally as a *chromophore*) attached to the opsin by covalent bonds. The chromophore is a derivative of retinol or vitamin A (Fig. 2.4), and exists in the fully dark adapted eye in the 11-*cis* stereochemical form. Upon exposure to light, the pigment is bleached (i.e., its specific absorptive properties are lost) and the chromophore changes rapidly to the all-*trans* form, and often, ultimately, into retinal aldehyde (Fig. 2.4). At the same time, one or more of the bonds attaching the retinol to the opsin are broken, leading, probably, to conformational changes in the three-dimensional structure of the opsin protein. The protein moieties are integrated into the matrix of the retinal receptor in an organized manner, and it is thought that the conformational changes in the opsin molecules lead to charge displacements which initiate the neural response.

The molecular changes described above involve several different intermediate forms of the visual pigment, and certain of the intermediates can be converted back to the original rhodopsin by light of a different wavelength. It will be seen in later chapters that phytochrome, the major pigment involved in photomorphogenesis in plants, has an impressively similar list of properties. It exists in two photoconvertible

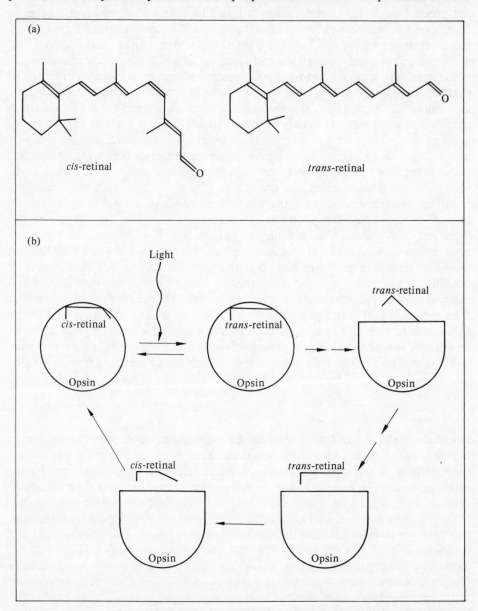

Fig. 2.4 (a) The chromophore of rhodopsin, the visual pigment of the human eye. This substance, known as retinal, is converted by actinic light from the 11-*cis* form (left) to the 11-*trans* form (right).

(b) The photoisomerization of retinal triggers complex conformational changes in the protein moiety of rhodopsin in which the chromophore becomes detached. Upon the reverse isomerization, reassociation occurs. These events bring about a charge separation in the disk membrane which evokes a neural signal (after Hendricks).[180]

forms, consists of a protein moiety and a photoreceptive chromophore, is probably organized into the membrane structure of the cells, and upon photoactivation undergoes electronic and configurational changes similar to those shown by rhodopsin.

2.5 Sources of light

There are three major types of light source:

1. the *thermal radiators*, which include the sun and incandescent lamps, and which have a continuous spectral distribution of energy with respect to wavelength, with a single maximum of energy per unit wavelength determined by the absolute temperature of the source;

2. the *discharge sources*, which include the arc lamps and discharges in metallic vapours, and in which the radiant energy is composed of lines characteristic of the emission spectra of the elements in the discharges, superimposed on a thermal background from the hot gases and electrodes; and

3. the *fluorescent lamps*, in which ultraviolet radiation produced by a low-pressure discharge is absorbed by a phosphor coating on the inside of the tubes, and then re-emitted as fluorescence at longer wavelengths.

(a) The black-body radiator

For our purposes, it is reasonable to assume that all thermal radiation sources are approximations to the ideal radiator of the physicist, known as the *black-body* radiator. The hypothetical black-body radiator is so-called because it is assumed to absorb all radiant energy of all wavelengths, and not to reflect or transmit radiant energy. For such a body, the radiation properties may be completely derived from thermodynamic and quantum theories, and the spectral energy distribution, and total flux radiated, are functions of temperature only. In practice, a black-body radiator can be closely approximated by a uniformly heated, opaque, usually metallic, enclosure or cavity, containing a small aperture for measuring the radiant energy. The radiant energy emitted by the walls of the cavity, is subjected to multiple internal reflection, and this energy is ultimately all absorbed by the walls, except for the small portion escaping through the sampling aperture.

The total flux per unit area (W) is proportional to the fourth power of the absolute temperature (T) for a black-body radiator as is stated by the Stefan–Boltzmann law:

$$W = \sigma T^4$$

where σ is a constant. Furthermore, the wavelength of maximum radiant energy emission per unit wavelength (λ_{max}) nm is also related to the absolute temperature (°K) in accordance with:

$$\lambda_{max} T = b$$

where b is a constant, 2.896×10^6 nmK. Thus, at a surface temperature of 2896°K, λ_{max} is 1000 nm, and the higher the temperature of the surface of the black-body source, the lower the wavelength of maximum emission.

(b) The sun

The sun behaves as a black-body source with a surface temperature of about 6000–6800°K; this temperature is maintained by nuclear reactions occurring within the body of the sun. The sun is the source of all visible light energy on the earth and variations in irradiance, spectral energy distribution, and daylength are of great importance in many photomorphogenic phenomena. As the sun can be considered as a black-body radiator with a temperature of 6000°K, the wavelength of maximum energy per unit wavelength, outside the earth's atmosphere, is about 470 nm, with a rapid decline on both sides of this peak (Fig. 2.5). However, as the sun's rays traverse the atmosphere, certain wavelength regions are selectively absorbed, in particular the ultraviolet (300–400 nm) absorbed by the ozone layer, and the infra-red, absorbed by water vapour (absorption bands between 720 nm and 2300 nm and complete absorption beyond). Throughout the visible region, clear air absorbs only about 20 per cent of the radiant energy when the sun is directly overhead, i.e., with an air mass of one.

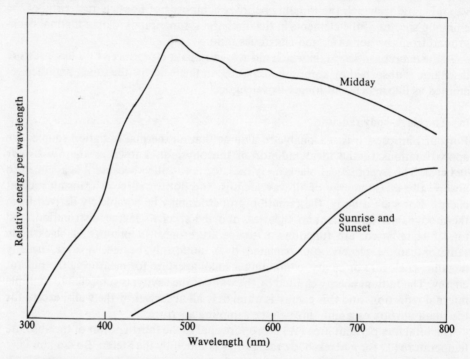

Fig. 2.5 The spectral distribution of radiant energy in direct sunlight at midday and at sunrise and sunset (from Smith).[402]

As the angle of incidence of the sun's rays increases (i.e., as the sun moves from overhead towards the horizon), the air mass value (i.e., the path length through the atmosphere) increases proportionally to the secant of the angle, and the energy is absorbed more strongly (Fig. 2.5). It is of particular interest to students of photo-morphogenesis that the wavelengths of maximum energy per unit wavelength shift from 470 nm at an air mass of one to about 650 nm at an air mass of five. Thus, both total flux intercepted at the earth's surface, and spectral energy distribution, are affected by the position of the sun as determined by the time of day, by latitude, and by season.

Other factors affecting incident irradiance are cloud cover, altitude, and pollution. The transmission of the atmosphere, on a heavily clouded day, may be only a few per cent of that on a clear day, while the values in high mountainous regions of dry areas of the earth may be more than 20 per cent greater than those at sea level.

The length of the day, or more strictly, the length of the night, is an important factor determining the morphogenic patterns of many plants, and also the behaviour of certain animals. Daylength, of course, does not vary a great deal at the equator, but as latitude increases, there is a consequent increase in the variation in daylength throughout the year. Maximum daylength in the northern hemisphere occurs on 21 June and minimum daylength on 21 December; these dates are obviously reversed for the southern hemisphere. The variations in daylength at different times of the year for different latitudes, go a long way to explaining the occurrence of photoperiodic ecotypes of many plants, in which the flowering and dormancy patterns have become adapted to the daylength regime of the growing regions (see Prue[343]).

(c) Artificial radiation sources

Most of the significant experiments in the study of photomorphogenesis have involved the use of artificial light sources, often with selected spectral regions isolated by various filters. The most commonly used primary light sources are the incandescent tungsten filament lamp and the fluorescent lamp.

The tungsten lamp is a thermal radiator in which the tungsten filament is heated by an electric current to 3000–3400°K, which is only a little below the melting point of tungsten at 3653°K. The spectral energy distribution varies with the type of lamp, but in all cases by far the greater proportion (up to 90 per cent) of the total flux emitted is in the infra-red region, with a λ_{max} usually around 1000 nm (Fig. 2.6).

In the fluorescent lamps, however, a low-pressure mercury discharge produces radiation at 253·7 nm, which is absorbed by the phosphor coating on the inside of the tube and re-emitted at higher wavelengths as fluorescence or phosphorescence. The delayed emission by phosphorescence is desirable since it damps out the flicker associated with the frequency of the discharge current which is noticeable with the rapid fluorescence phosphors. White fluorescent lamps have mixtures of two or more phosphors emitting different wavelengths, and by using various combinations of different phosphors various 'colours' of white light can be obtained, e.g., the 'warm' white, 'cool' white and, 'daylight' tubes (Fig. 2.6). A major advantage of fluorescent lamps is that the radiant energy is almost all within the visible range (Fig. 2.6), and thus calculations of total effective irradiance are more easily made. Another specific advantage for photomorphogenic studies is the low content of radiant energy above 700 nm in this lamp, allowing the simple construction of red light sources, containing no radiation in the 720–760 nm region (now known as the *far-red*), which as will be seen later is a particularly important region of the spectrum (see Appendix for details of the construction of some simple sources).

The isolation of specific wavelength regions from these primary sources is normally achieved by the use of filters of varying types, the two major types in use at present being the broad-band filters, and the narrow-band interference filters. Broad-band filters isolate, as the name suggests, broad spectral regions and can be composed of several types of material. Early filters usually consisted of solutions of specific, coloured, inorganic, or organic substances of known absorption properties, or of thin films of gelatine into which known concentrations of specific dyes had been incorporated, and which were sandwiched between glass sheets. In later years, several glass companies have produced very wide ranges of coloured glass filters of

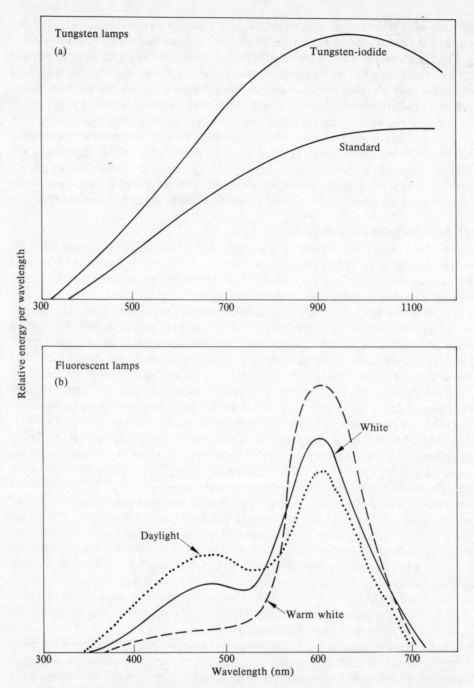

Fig. 2.6 The spectral distribution of radiant energy in (a) tungsten incandescent lamps, and (b) white fluorescent lamps of different types. In (1) the emission lines from the gas in the tube are omitted (from Canham).[67]

very rigorously controlled specifications, and most recent of all has been the introduction of coloured plastic sheeting used principally in the theatrical industry for spotlights and other lighting effects. These broad-band filters can be combined to produce filter systems which isolate relatively narrow bands of radiation, and have been extensively used in photomorphogenic experiments.

Light which is almost monochromatic, consisting of a very narrow band of wavelengths, can be isolated by interference filters, which consist normally of two or more sheets of semi-transparent reflective material sandwiching a transparent film of a thickness equal to half the wavelength to be transmitted. This wavelength is then selectively transmitted by constructive interference within the sandwich. These filters are, naturally, very much more expensive than the broad-band filters and large areas cannot be produced. Furthermore, the incident radiation must be collimated to within an angle of incidence of less than 7°, or the effective distance between the sandwich increases, and other wavelengths are transmitted. Monochromatic light is essential for the construction of action spectra and it is in such experiments that interference filters have found their widest use.

Action spectra have also been worked out using other methods of isolating restricted wavelength regions, the most spectacular being the spectrograph used at the United States Department of Agriculture Experimental Station at Beltsville, Maryland. In this instrument, two very large glass prisms were used to split the light from a powerful carbon arc source into a spectrum which was projected onto an experimental area at a total width of up to two metres. Experimental plants were then spread out along this spectrum and irradiated with known energies of known wavelengths. These experiments have been largely responsible for one of the most important developments in the study of photomorphogenesis in plants, as is described in chapter 3. (Simple experimental light sources are described in the Appendix.)

3

The photomorphogenic response systems and their photoreceptors

We are concerned here only with those developmental processes which are regulated directly by non-periodic, non-directional light stimuli. Within this field, it is possible to distinguish at least three photoreactive response systems which differ, either in the nature of the photoreceptor used, or in the energy levels required to elicit the responses. It is the function of this chapter to describe the known response systems, and to identify, if possible, the photoreceptors responsible for initiating the responses. Although directional effects of light do not strictly fall into the field of coverage, it is, nevertheless, instructive to consider the energy relationships and the nature of the photoreceptor for phototropism since, as will be seen, it seems likely that the phototropic photoreceptor is identical with one of those involved in photomorphogenesis.

Photomorphogenic phenomena can be found at all stages of plant development, but the most intensive experimentation has been devoted to two major processes – seed germination, and the recovery from etiolation (otherwise known as de-etiolation). That there should have been such a concentration of effort on a restricted range of phenomena is to some extent unfortunate, since photomorphogenesis is intimately involved at all stages of the regulation of plant development in the natural environment, and is not restricted to the developmental processes of germinating seeds, and of seedlings grown unnaturally in darkness. On the other hand, the physiological investigations into the ways in which light regulates seed germination, and de-etiolation, have led to remarkable advances in our understanding of the subject. Furthermore, they constitute an object lesson in scientific method which is without parallel in other areas of plant physiology.

3.1 The red/far-red photoreversible response system

Between 1935 and 1937, two observations were reported which can be said to mark the beginning of the analytical study of photomorphogenesis. In these publications, it was reported that the effects of red light (i.e., *ca.* 600–700 nm) and near infra-red light (i.e., *ca.* 700–800 nm) on seed germination were in certain circumstances opposite in direction. The authors, Flint and McAllister[120, 121] showed that a certain variety of lettuce seed could be stimulated to germinate by red light treatment, and that the endogenous germination in the dark, which was already low (*ca.* 30 per cent), was even further inhibited by near infra-red light. These responses required only very low energies of irradiation. These observations confirmed several earlier reports of developmental effects initiated by treatment of plants with minute amounts of red and near infra-red light. Thus, a picture began to emerge of a low-energy photo-response with action maxima in the red, and the near infra-red, spectral regions. There were also indications that the effects of red and near infra-red light could be opposite in direction.

In the 'forties, this work was taken up by a group of investigators based at the United States Department of Agriculture Research Station at Beltsville, Maryland, in what was to become a lifetime's study of photomorphogenesis. The group was led by Dr H. A. Borthwick, a botanist, and Dr S. B. Hendricks, a physical chemist, two names which will always be associated with the most important discoveries in the field. Their early aims were to construct detailed action spectra of the known photomorphogenic phenomena, and to this end they constructed the spectrograph described above (page 21).

With this equipment they produced accurate action spectra for the effects of light on many developmental processes, including the photoperiodic induction of flowering in *Xanthium*,[323] control of stem growth in etiolated barley seedlings,[36] control of stem and leaf growth in etiolated pea seedlings,[324] and the control of germination in seeds of the Grand Rapids variety of lettuce.[39] Representative action spectra are shown in Fig. 3.1, from which it is clear that the two action maxima occur at, very nearly, the same wavelengths in all the responses, and that light of the two wavelengths, in each case, gave opposite results. The precise wavelengths of maximum action were 660 nm in the red, and 730 nm in the far-red region. (The spectral region between *ca*. 700 nm and *ca*. 800 nm is now known as the *far-red* by plant physiologists.)

It was therefore demonstrated that a low-energy photoresponse, with opposite actions in the red and far-red regions of the spectrum, was widespread in flowering plants. What is now recognized as being probably the most important experiment in the history of photomorphogenesis was then performed. Borthwick, Hendricks and their associates showed in 1952 that the responses to the two wavelengths were not merely opposite; they were also antagonistic.[37] They found that lettuce seeds potentiated to germinate by red light would not germinate if the red light treatment was quickly followed with a far-red light treatment. Furthermore, if a series of alternating red and far-red light treatments were applied, the ultimate germination was determined by the nature of the last treatment; if it was red light, then a high incidence of germination would result, if far-red light, then a low incidence of germination would result. The results of this now classic experiment are given in Table 3.1.

Table 3.1 Control of lettuce seed germination by red and far-red light (from Borthwick *et al*.[37]).

Irradiation	Percentage germination
Red	70
Red/far-red	6
Red/far-red/red	74
Red/far-red/red/far-red	6
Red/far-red/red/far-red/red	76
Red/far-red/red/far-red/red/far-red	7
Red/far-red/red/far-red/red/far-red/red	81
Red/far-red/red/far-red/red/far-red/red/far-red	7

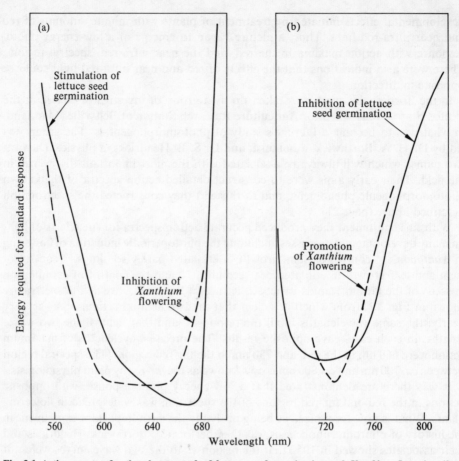

Fig. 3.1 Action spectra for the photocontrol of lettuce seed germination and *Xanthium* flowering (from Borthwick).[34]

With these results, Borthwick and Hendricks were able to make some far-reaching predictions concerning the nature of the photoreceptor for this red/far-red photo-reversible response. They argued that the physiological response pattern observed could be most simply explained by assuming that the photoreceptor existed in two interconvertible forms, one of which had maximal absorption at 660 nm, and the other at 730 nm. Absorption of light by either form would result in conversion to the other form. This hypothesis can be represented as follows:

$$\text{Pr} \underset{\text{730 nm}}{\overset{\text{660 nm}}{\rightleftharpoons}} \text{Pfr}$$

where Pr and Pfr represent the two forms of the pigment P absorbing maximally at 660 nm in the red, and 730 nm in the far-red, respectively.

The next obvious step was to obtain some more direct evidence for the existence, in plant cells, of a pigment with the predicted properties. The problems, however, were immense. The primary biological action of the pigment was unknown (as it still largely remains) and it was therefore not possible to search for it by means of any enzymatic or biological assay. The only possibility was *in vivo* spectrophotometry; i.e., to attempt to detect light-mediated changes in the absorbance of plant tissues at

24   Phytochrome and photomorphogenesis

the wavelengths of maximum absorption by the postulated pigment, i.e., 660 nm and 730 nm.

The two major difficulties attending the *in vivo* spectrophotometric detection of the postulated photoreversible pigment were the overriding absorbance of chlorophyll in green tissues, and the design of a suitable instrument. These problems were eventually overcome by the use of etiolated tissue, and a specially constructed spectrophotometer capable of recording the absorbance of very dense, light-scattering samples.[65] The first tissue to be used was the cotyledons of dark-grown turnip seedlings, grown in the presence of the antibiotic chloramphenicol, a substance which specifically inhibits protein synthesis in cell organelles, such as plastids. This treatment resulted in a tissue with a very low protochlorophyll content, a distinct advantage

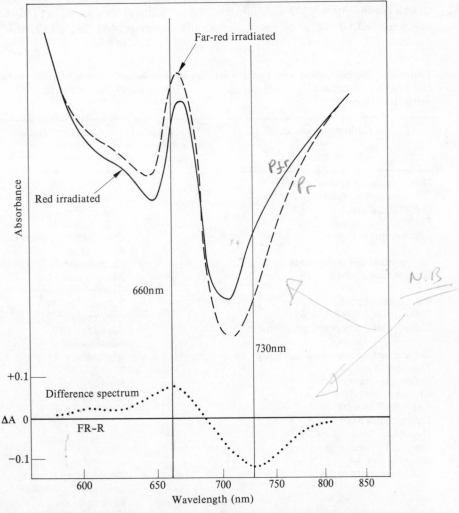

Fig. 3.2 The first direct demonstration of phytochrome by *in vivo* spectrophotometry. The upper curves show the actual absorption spectra of etiolated maize shoots after irradiation with actinic sources of red and far-red light. The large absorption peak is due to protochlorophyll. The lower difference spectrum, calculated by subtracting the 'red' absorption spectrum from the 'far-red' absorption spectrum, shows that red light increases the absorbance of the tissue at 730 nm while decreasing it at 660 nm (from Butler *et al.*).[65]

since red light treatment irreversibly converts protochlorophyll to chlorophyll, thus contributing absorbance changes which are not related to the photomorphogenic pigment. The experimental procedure was to irradiate the etiolated tissue in a cuvette with red light, and then to determine the visible absorption spectrum of the tissue. This procedure was then repeated, but the actinic irradiation used was far-red light. Typical results, obtained in this case with etiolated maize seedlings, are shown in Fig. 3.2, demonstrating that red light causes an absorption decrease in the 660 nm region and an increase in the 735 nm region, while far-red light reversed these spectral changes, thus elegantly verifying the predictions based on the physiological data. The authors corroborated this finding by showing that buffered aqueous extracts of etiolated tissue also contained a substance which gave similar spectral changes on irradiation with red and far-red light. This substance was partially purified and shown to be a chromoprotein. This elegant piece of work therefore showed not only that the predicted red/far-red photoreversible photoreceptor existed, but also that it was a protein.

Table 3.2 Distribution and nature of the low-energy phytochrome system in the plant kingdom (+ = promotion, − = inhibition) (after Furuya,[140] with additions).

Photoresponse	Direction of red light action	Plants studied
Non-vascular lower plants		
Oospore germination	+	*Chara*[413a]
Cell elongation	+	*Chara*[353a]
Filament branching	+	*Mougeotia*[315a]
Spore germination	+	*Ceratodon*[442], *Dicranum*[442], *Funaria*[11]
Cell number per protonema	+	*Ceratodon*[442], *Dicranum*[442],
Growth rate of thalli	+	*Marchantia*[127], *Sphaerocarpos*[286]
Chloroplast (size)	−	*Polytrichum*[165]
Chloroplast (division rate)	+	*Polytrichum*[165]
Chloroplast (phototactic movement)	+	*Mougeotia*[169], *Mesotaenium*[176]
Chloroplast (chlorophyll content)	+	*Marchantia*[127]
Pteridophyta		
Spore germination	+	*Osmunda*[303]
Growth (rhizoid)	+	*Onoclea*[298]
Growth (filament)	−	*Onoclea*[288]
Gymnospermae		
Hypocotyl hook formation	+	*Picea*[375]
Longitudinal stem growth	−	*Picea*[375]
Angiospermae		
Seed germination	+	*Anagallis*[153], *Chenopodium*[79], *Lactuca*[37], *Lycopersicon*[275], *Paulownia*[41].
Seed germination	−	*Eragrostis*[130,436]
Growth rate (plant height)	−	*Pisum*[265]
Growth rate (final leaf number)	−	*Secale*[128]
Growth rate (intact leaf length)	+	*Phaseolus*[89], *Pisum*[135,455]

Table 3.2 (continued)

Photoresponse	Direction of red light action	Plants studied
Growth rate (leaf disc size)	+	Phaseolus[262, 334, 335]
Growth rate (intact coleoptile)	−	Oryza[331]
Growth rate (excised coleoptile)	+	Avena[199, 261]
Growth rate (intact mesocotyl)	−	Avena[268]
Growth rate (intact stem length)	−	Phaseolus[89, 91], Pisum[430]
Growth rate (intact stem length)	+	Pisum[430]
Growth rate (sectioned stem)	−	Pisum[15, 16, 191]
Frond size and root length	+	Lemna[223]
Hook opening	+	Phaseolus[303], Cuscuta[255]
Hyponastic leaflet movement	+	Hyoscyamus[378]
Closing leaflet movement	+	Mimosa[122], Albizzia[195]
Geotropic reactivity	+	Avena[23, 458], Sinapis[304]
Geotropic reactivity	−	Zea[459]
Phototropic reactivity	−	Zea[46]
Photoperiodism (light break effects)		
Photoperiodism short-day plants	−	Xanthium[90], Pharbitis[314, 414], Lemna[189]
Photoperiodism long-day plants	+	Chenopodium[80], Hordeum[90], Hyoscyamus[90]
Stabilization of vernalized condition	+	Secale[128]
Auxin-induced lateral root initiation	+	Vigna[129]
Auxin-induced lateral root initiation	−	Pisum[23]
Leaf primordia formation	+	Sinapis[305]
Cytoplasmic viscosity	+	Kalanchoe[384, 462]
Dry weight (leaf)	+	Pisum[135, 186], Zea[73]
Protein content (seedling)	+	Sinapis[254]
Transport of sucrose	+	Pisum[148]
Starch/sugar degradation	+	Zea[245, 291, 338, 339]
Anthocyanin biosynthesis	+	Brassica[157], Sinapis[292]
Flavonoid biosynthesis	+	Pisum[43, 135, 310, 365, 404]
Carotenoid biosynthesis	+	Pisum[186], Zea[73]
Chlorophyll content	+	Pisum[186]
Ascorbic acid content	+	Fragaria[280]
ATP concentration	+	Avena[394], Phaseolus[394]
Glyceraldehyde-3-phosphate dehydrogenase activity	+	Phaseolus[277]
Amino acid activating enzyme activity	+	Pisum[186]
Phenylalanine ammonia-lyase formation and activation	+	Pisum[7], Sinapis[97], Fagopyrum[377]
Change in electric potentials	+	Phaseolus[210], Avena[315]
Adhesion of roots to glass	+	Hordeum[416], Phaseolus[415]

Shortly after these results were published, Borthwick and Hendricks[34] coined the term *phytochrome* (*phyto* ≡ plant, *chrome* ≡ colour, pigment) for the photoreceptor.

Thus, it was discovered that one of the ways in which light could directly regulate the development of plants was through the absorption of small quantities of light by the proteinaceous photoreceptor, phytochrome. This category of response is characterized by saturation at low energies of irradiance, and short times of irradiation, and by the antagonistic and reversible effects of red and far-red light.

3.2 The typical low-energy phytochrome response pattern

The reversible effects of red and far-red light on germination described above are now known to occur in an extremely wide range of species, covering a similarly wide range of developmental processes. Such reversible responses to red and far-red light of low intensity have been reported to occur in species ranging from the single celled algae to the flowering plants, although they have not yet been found in any fungus. The types of response vary from effects on flowering, which may take days or weeks to be manifested, to effects on leaflet closing and electrical potential changes, which become measurable within 15 to 30 seconds. Table 3.2 presents a comprehensive list of the major categories of developmental and metabolic responses that are known to be regulated by the low-energy red/far-red photoreversible reaction.

This response pattern is now known as the 'phytochrome system', or the 'phytochrome reaction'. As will be seen, however, phytochrome is almost certainly involved in other photoresponses which require higher energies, and thus such a name is apt to be ambiguous. The low-energy red/far-red reversible reaction will therefore be referred to here as the *low-energy phytochrome system*.

To determine whether a certain photomorphogenic phenomenon is mediated through the low-energy phytochrome system, it is necessary to apply certain criteria as follows:

1. Is the process red/far-red reversible?
2. Are the response maxima located at or near 660 nm and 730 nm?
3. Is the response saturated at low energies?

Satisfaction of all these criteria can be taken as unequivocal proof of the involvement of the low-energy phytochrome system, although the simultaneous operation of other systems is not thereby ruled out.

These diagnostic tests stem from the realization that the low-energy phytochrome system normally manifests itself through a common response pattern that is generally similar in all cases. Variations to this typical response pattern must be expected, since we are dealing with the diversity of living organisms, but the response pattern is sufficiently general for it to be considered under a number of headings.

(a) Direction of red light action

Irradiation of an etiolated pea plant with red light causes the stimulation of leaf expansion and the overall inhibition of stem elongation.[324] Red light, therefore, operates in this case to stimulate one developmental process while inhibiting another. This phenomenon is widespread and cases are known of red light stimulating a process in one plant, while inhibiting the very same process in a closely related plant. Root initiation is a good example of this diversity of response. In *Vigna sesquipedalis*, red light promotes lateral root initiation,[129] yet it inhibits root initiation in the closely related species *Pisum sativum*.[156] In several other responses, the direction of the red effect changes with the age of the plant. In *Pisum sativum*, again, red light promotes the elongation of young internodes, but inhibits that of older internodes.[430] Other physiological treatments can also alter the direction of the response to red light. The elongation of intact oat coleoptiles is inhibited by red light, but if apical sections of the coleoptiles are excised from the plant and floated on buffered sucrose solution, the red light causes a marked stimulation of growth.[200]

These diverse responses cannot be understood in terms of qualitatively different primary mechanisms of the photoreceptor. They are more likely caused by the modifying actions of some metabolic processes which interact with the photoreceptor, and thus determine the final result.

(b) Energy required for saturation

One of the major criteria described in section 3.2 is the low-energy requirement. In the literature, the incident red energy required for saturation of photoresponses falls mainly within the 1–1000 Joules m^{-2} range. Some typical values are given in Table 3.3. Much of this energy, of course, will be lost to the phytochrome system through reflection, light-scattering within the tissues, and through absorption by other coloured substances, especially chlorophyll and protochlorophyll. It is thus not possible to compare quantitatively responses manifested in different plants. In any case, the values are very low, and preclude the possibility that phytochrome is operating as an energy transducer in a similar manner to chlorophyll. The values are much more in line with a photoactivation process, in which only a small amount of energy is required for the formation of a biologically active component.

Below the saturation level, but above a certain, usually very low, threshold value, the responses to red light have been found to be typically directly proportional to the logarithm of the incident energy. Two characteristic examples are given in Fig. 3.3 which shows the increase in the activity of phenylalanine ammonia-lyase, an enzyme involved in secondary product synthesis, in pea seedlings, as a function of red light energy,[7] and the increase in barley leaf width as a function of red light energy.[399] The logarithmic relationships are shown clearly. Many other similar examples have been reported, such as the growth of pea leaves,[135] leaf weight[247] and anthocyanin synthesis,[247] and bean hook opening and hypocotyl growth.[464]

Table 3.3 The red light energy required to produce maximum developmental responses (after Furuya[131]).

Photoresponse	Saturation energy Joules m^{-2}	Materials and references
Epicotyl hook opening	0·1	Phaseolus[246]
Epicotyl hook opening	1·0	Cuscuta[255]
Geotropic reactivity	1·0	Avena[458]
Seed germination	10·0	Lactuca[463]
Coleoptile elongation	10·0	Oryza[331]
Plumular growth	30·0	Pisum[135]
Inhibition of flowering by light break	30·0	Xanthium[435]
Stem section growth	60·0	Pisum[16]
Mescocotyl growth	300·0	Avena[268]
Inhibition of flowering by light break	1980·0	Pharbitis[414]

Note. The original data have been converted to SI units.

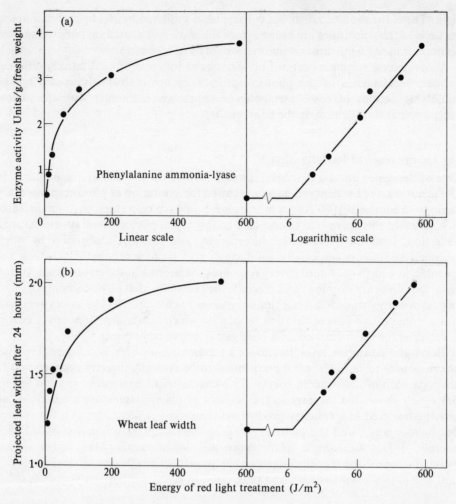

Fig. 3.3 Two examples of the logarithmic relationship between red light energy and photomorphogenic response:

(a) the red-light stimulated increase in the enzyme phenylalanine ammonia-lyase in pea seedlings (from Attridge and Smith),[7] and

(b) the red-light stimulated unrolling of the primary leaf of etiolated barley seedlings (unpublished results).

One of the early conclusions made from the very low energies required for substantial responses was that the form produced by red light was the biologically active form. The response to red light could, in theory, be due either to the conversion of an active inhibitory form to an inactive form, or to the conversion of an inactive form to an active, stimulating form. The logarithmic relationship between incident energy and ultimate developmental response suggested that the photoact was not removing an active inhibitory form. Such a process would show a markedly different dose-response relationship. Furthermore, as is described in detail in chapter 5, it was found that the red light inhibition of pea epicotyl extension was saturated when only 50 per cent of total phytochrome had been photoconverted.[191, 192] Thus, it was concluded that the red-absorbing form of phytochrome was biologically inactive, and that the formation of small amounts of the far-red-absorbing form would lead to some measure of response, even in the presence of proportionately large amounts of the red-absorbing form. This conclusion fits the facts, but ignores a third possibility as regards the active form of phytochrome; namely, that neither form of phytochrome is itself responsible for the ultimate morphogenic processes, but that those processes may occur as the result of some metabolic step that is started by the act of photoconversion. Using this hypothesis, it is the actual process of photoconversion that is important, not the new form of phytochrome produced by the photoconversion. Consideration of this idea will be postponed until chapter 5, and in the meantime we will use the more generally accepted view that red light causes an activation of phytochrome.

In most cases, the energies required for maximum reversal by far-red light are three to four times greater than the energies of red light required for maximum response. This could be due to one or more of several different possibilities:

1. that the plant tissues may exert a greater screening effect at the far-red wavelength regions; or
2. phytochrome in its Pfr form may have a lower extinction coefficient than the Pr form; or
3. that the quantum efficiency of the Pr to Pfr conversion may be greater than the quantum efficiency in the reverse direction.

To answer these questions, knowledge of the photochemical properties of phytochrome itself is required, and consequently further consideration of this subject will be postponed until chapter 4.

(c) Time-course of phytochrome action

The pattern that emerges therefore is that red light converts the inactive red-absorbing form of phytochrome to the active far-red-absorbing form, which then mediates the ultimate metabolic or developmental events, presumably via some amplifying mechanism. This process can be reversed by immediate photoconversion in the opposite direction by far-red light. The overall reaction is commonly written:

$$Pr \underset{\text{far-red light}}{\overset{\text{red light}}{\rightleftharpoons}} Pfr \longrightarrow \text{biological action}$$

where Pr is the red-absorbing form, and Pfr the far-red-absorbing form. In the literature, Pr is sometimes known as P_{660}, and Pfr as P_{730}. Both conventions are acceptable although Pr and Pfr is more common.

The immediate metabolic effects of the photoconversion of Pr to Pfr are, as yet, unknown, and it is quite clear that the ultimate developmental events are quite remote from these initial processes. In all cases already described a detectable lag phase occurs between the onset of irradiation and the earliest detectable responses. In most morphogenic responses, the first changes usually become measurable only several hours after the photoactivation of phytochrome. Certain other red light-initiated responses which do not involve cellular growth and development, such as leaflet movement and chloroplast reorientation, are known to occur with a much shorter lag phase, measurable in minutes only. Some characteristic lag phases are given in Table 3.4. Even for some of the responses which are not detectable for many hours (e.g., seed germination), it is possible to show by ingenious physiological experiments that the action of phytochrome is completed very rapidly after light treatment. These experiments are considered in detail in chapter 5 since they provide valuable information on the mechanism of action of phytochrome.

Table 3.4 Comparison of lag periods between red light irradiation and the onset of a detectable response (from Furuya[131], with later additions).

Criterion studied	Lag period	Plant material and references
Change in electric potentials	15 s	Oat coleoptiles[315]
	30 s	Mung bean root tips[210]
Change in adhesion of roots to glass	30 s	Barley[416]
Chloroplast reorientation	5 min	*Mougeotia*[171]
Closing movement of leaflet	5 min	*Mimosa*[122], *Albizzia*[195]
Increased gibberellin levels	10 min	Wheat leaves[12]
Increased rate of geotropic response	30 min	Oat coleoptiles[458]
Ascorbic acid synthesis	1 h	Mustard hypocotyls[379]
Decreasing phototropic sensitivity	1·5 h	Maize coleoptiles[48]
Increase in phenylalanine ammonia-lyase activity	1·5 h	Pea buds[403]
Inhibition of stem growth	1·5 h	Peas[264]
Increased proportions of polysomes	2 h	Bean leaves[329]
Increased rate of carbohydrate metabolism		Bean leaves[245]
Changes in flavonoid synthesis	2 h	Pea buds[404]
Increase in leaf growth rate	4 h	Peas[135]
Inhibition of stem section growth	4 h	Peas[16]
Decreased rate of geotropic response	4 h	Maize coleoptiles[459]
Increase in anthocyanin level	4–16 h	Mustard hypocotyls[294]

(d) Escape from photoreversibility

Insertion of a dark period between red and far-red irradiation treatments causes a partial failure of reversibility of the response being measured, the degree of reversibility remaining being dependent on the length of the dark period inserted. Measurement of the time-course of this escape from photoreversibility should provide valuable information on the rates of the intermediate processes of the photoactivated form of phytochrome. The observed rates of escape from photoreversibility, however, vary considerably between different species and between different processes. Obviously, in those cases where the red light-initiated display occurs rapidly after light treatment, then the time-course of the escape from photoreversibility is at least as rapid, but it is surprising that there is such wide variation in escape curves for the longer term morphogenic displays. This variation is shown in Table 3.5 which gives a list of the times taken for 50 per cent loss of photoreversibility for various processes. A typical escape curve, in this case for the red light-initiated unrolling of etiolated barley leaf sections, is given in Fig. 3.4.

On the basis of the simple model of phytochrome action on page 24, escape from photoreversibility must occur when the initial processes triggered, or catalysed, by Pfr have been completed. Only then would reconversion of Pfr to Pr have no effect on the subsequent steps. On the other hand, the electrical responses to phytochrome conversion described later (page 161) exhibit far-red reversal of the response itself, rather than reversal of the induction of the response.

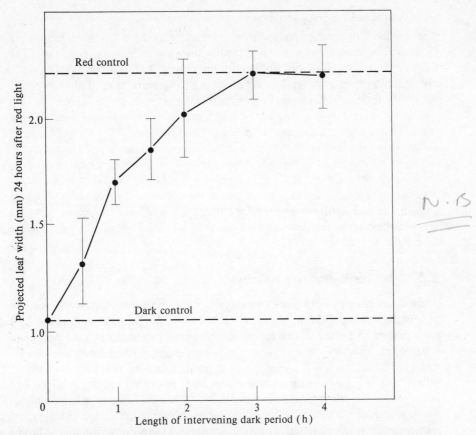

Fig. 3.4 The escape from photoreversibility of the barley first leaf unrolling process – the curve shows the data from an actual experiment in which barley seedlings were given red light 600 J m^{-2}, followed by far-red light, 1800 J m^{-2}, after various time intervals; 24 hours after the red light treatment projected leaf widths were measured. The vertical bars represent the standard errors. Clearly, far-red reversibility is completely lost 2·5 to 4 hours after the exposure to red light.

Table 3.5 Rates of escape from photoreversibility (after Furuya[131], with additions).

Process	Time for 50 per cent loss of photo-reversibility	Plant material
Light break in photoperiodism	1·5 min	*Pharbitis*[125], *Kalanchoe*[126]
	15 min	*Xanthium*[90]
	25 min	*Glycine*[90]
Flavonoid synthesis (IAA-oxidase inhibitor)	30 min	Peas[194]
Leaf unrolling	1 h	Barley (see Fig. 3.4)
Hook opening	4 h	*Cuscuta*[255]
Coleoptile growth (section)	6 h	Oats[201]
Coleoptile growth (intact)	8 h	Rice[331]
Seed germination	9 h	Lettuce[39]
Seed germination in gibberellin	5 min	Lettuce[17]

(e) Atypical red/far-red phenomena

The typical low-energy phytochrome response pattern outlined here is, like all biological processes, subject to wide variations, particularly in the timing of the light treatments required to obtain maximum effects. One of the commonest of the atypical responses is the case in which a single exposure to red light has no effect, or only very little effect, while if the same total energy is given in brief flashes every one or two hours over a long time period, maximum effect is achieved. This type of response has been investigated in *Sinapis*[292] and *Cucumis*[196] seedlings, where it was shown that the effects of the intermittent flashes of red light could be reversed if each red light treatment was immediately followed by treatment with far-red light. In these cases, intermittent red light given over a 20 hour period, for instance, has an equivalent, and sometimes greater, effect than that of a continuous exposure over 20 hours to red light – it seems likely that such responses involve a further photomorphogenic response system, the so-called high-energy reaction, which will now be described.

3.3　The high-energy reaction

The low-energy phytochrome system as described in the previous sections is saturated by low-light energies given over short time periods. Such conditions are not frequently encountered in nature where plants are commonly exposed to long periods of high irradiances, and the existence of other photoreactions responding to natural conditions of illumination is therefore to be expected. At least two photoreactions falling into this category can at present be recognized, both having been grouped under the collective term, '*the high-energy reaction*'. This term was originally suggested by Mohr in 1957[292] to distinguish these photoreactions from the low-energy, red/far-red photoreversible photoreaction. Other workers[445] prefer to designate these photoreactions as '*prolonged light reactions*' to draw attention to the fact that in most cases reciprocity does not hold, and the absolute length of the period of illumination is more important than the total incident quanta, in determining the magnitude of the biological response. For the purpose of this discussion, the term 'high-energy reaction' will be used, mainly because it is prevalent in the literature, but also because it emphasizes the important fact that these reactions are dependent on high irradiances.

The original distinction between the low-energy, red/far-red photoreversible (phytochrome) reaction and the high-energy reaction was made by Mohr working on cotyledon expansion in mustard (*Sinapis alba*) seedlings.[292] If fully dark-grown seedlings are given short daily periods of red light the cotyledons, which do not normally grow in darkness, expand, this response being saturated at low energies and short times of irradiance. This stimulation of cotyledon expansion can be negated by exposing the seedlings to short periods of far-red light immediately after each daily red light treatment, showing that phytochrome is involved, but if the period of far-red light is increased beyond approximately 2–3 hours no reversal is seen, and in fact a stimulation of expansion occurs. It did not seem possible at this stage of understanding to reconcile this pattern of events with the low-energy, red/far-red photoreversible system, and Mohr therefore postulated the existence of a further photoreceptive pigment involved in the regulation of plant development which required long periods of relatively high irradiances for maximum effect.

The high-energy reaction has subsequently been implicated in a very wide range of

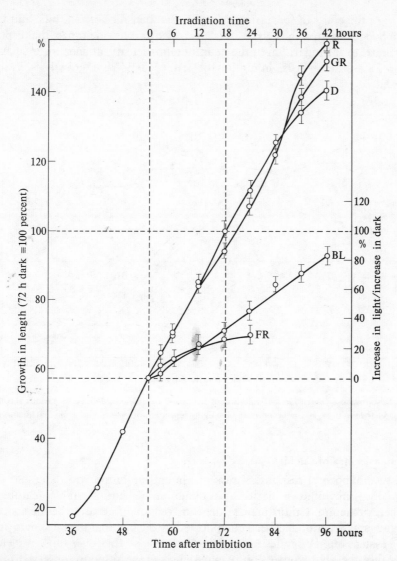

Fig. 3.5 The effects of a sustained irradiation with various wavelengths of light on hypocotyl extension in otherwise dark-grown lettuce seedlings. Both blue (BL) and far-red (FR) light inhibit growth, but far-red is most effective (reprinted with permission from Hartmann).[167]

developmental phenomena, as wide a range, in fact, as that in which the red/far-red phytochrome system is involved, and including seed germination,[299] stem[307] and leaf[295] expansion, anthocyanin[292] and flavonol[404] synthesis, nucleic acid[455] and protein[452] synthesis, and changes in the levels of specific enzymes.[97, 320]

A typical example of a high-energy reaction is the inhibition of hypocotyl elongation in lettuce (*Lactuca sativa*) seedlings.[307] This response has been analysed in great detail by Hartmann, and the time-course of the effect with different wavelengths of light is shown in Fig. 3.5. It can be seen that only blue and far-red light cause any significant inhibition of growth. In this experiment, all the spectral bands were used at the same quantum flux density so that meaningful comparisons could be made.

In Fig. 3.6, the effect of increasing quantum flux densities of red, blue and far-red light on hypocotyl growth is given. It is clear that over a wide range, the response is proportional to the logarithm of the quantum flux density, at least for blue and far-red light. In other words, inhibition of lettuce hypocotyl growth is irradiance dependent.

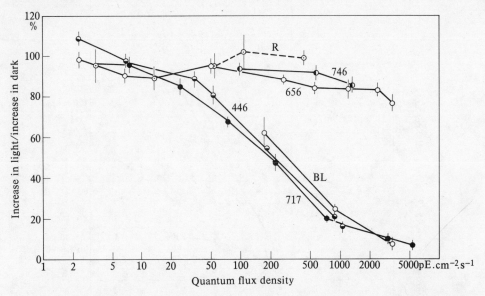

Fig. 3.6 Dose-response curves for the inhibition of hypocotyl extension in lettuce by various wavelengths. R = broad band red light, BL = broad band blue light (reprinted with permission from Hartmann).[167]

(a) Action spectra of the high-energy reaction

The lettuce hypocotyl response shows maximum inhibition with blue and far-red light, and it is now known that this is a common, but not inevitable, feature of the high-energy reaction. Unfortunately, the construction of accurate action spectra for the high-energy reaction has been complicated by the fact that the low-energy phytochrome system often operates in the same response so that over those wavelengths which cause the photoconversion of phytochrome the response observed may not be due solely to the high-energy reaction. In most developmental responses regulated by the high-energy reaction, the response, if any, to low-energy phytochrome photo-activation is in the same direction as, but usually much smaller than, the response to high energies of continuous irradiation. Thus, the responses mediated by the two systems in these cases can be expected to be additive, if they operate through different mechanisms. Using this assumption, Mohr[293] was able to obtain an action spectrum for the expansion of mustard seedling cotyledons by the following procedure. The response to a 4 hour period of monochromatic light at various irradiances was first determined. Then, the response to a number of flashes of that light, separated by darkness (which was considered to represent the response due to the low-energy phytochrome reaction alone), was subtracted from the values for the 4 hour treatments, thus, hopefully, giving the responses due to the high-energy reaction alone. This first action spectrum for the high-energy reaction (Fig. 3.7) shows a sharp maximum in the far-red at approximately 730 nm, with a low shoulder in the red at 630 nm, and a smaller peak at 440 nm with a shoulder at about 480 nm. Thus, there are two main wavelength regions of high quantum efficiency, the far-red and the blue. In Fig. 3.7 the action spectrum for short irradiations is included for comparison.

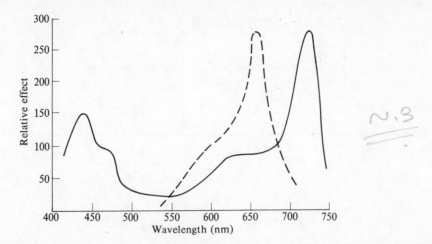

Fig. 3.7 A generalized action spectrum of the high-energy reaction (solid line) together with that of the low-energy phytochrome reaction (dotted line) for comparison (from Mohr[293] with permission from Biological Reviews).

A small number of other photomorphogenic responses controlled by both the low-energy phytochrome system and the high-energy reaction are affected in different directions by the two photosystems. For example, etiolated lettuce seedlings do not produce the plumular hook characteristic of most dark-grown dicotyledonous seedlings, but the formation of such a hook by asymmetric growth on the opposite sides of the hypocotyl can be induced by a brief treatment with low-energy red light. This induction with red light can be prevented by immediate subsequent irradiation with far-red light, thus proving that the formation of the hook is a classical low-energy phytochrome response. If, however, the hook is allowed to form in darkness after red light treatment, it can later be removed by continuous high-energy irradiation with either blue or far-red light. Since the low-energy phytochrome reaction and the high-energy reaction operate antagonistically in this response, there is less difficulty in constructing an action spectrum, which, in fact, exhibits similar peaks in the blue and far-red regions as that for mustard cotyledon expansion.[302]

The biosynthesis of anthocyanins in many plants is regulated by light in a manner which suggests the involvement of the high-energy reaction. Figure 3.8 provides a comparison of the action spectra for anthocyanin synthesis for four dicotyledonous plants and one monocotyledonous plant (*Sorghum*) illustrating the range of spectra obtainable. In four of the five cases, maximum effect is achieved in the far-red or red regions of the spectrum and three of these spectra show considerable response in the blue. The precise maxima range from 660 nm for apple fruit skins, to 730 nm for mustard and turnip, with red cabbage in between at approximately 705 nm. Only *Sorghum* seedlings do not respond to wavelengths in the red or far-red region, and in this case a single maximum in the blue is evident.

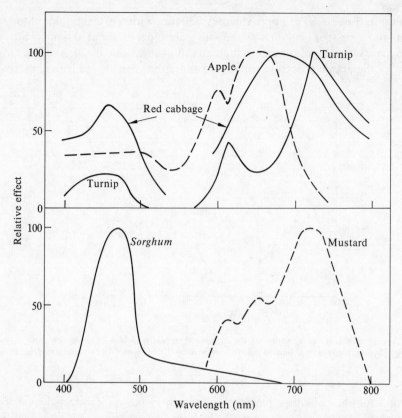

Fig. 3.8 Action spectra for the high-energy photocontrol of anthocyanin synthesis in several species. Note the great variation in wavelengths of maximum action. Redrawn from the published data; apple (from Siegelman and Hendricks)[388]; red cabbage, turnip (from Siegelman and Hendricks)[389]; mustard (from Mohr)[292]; *Sorghum* (from Downs and Siegelman).[92]

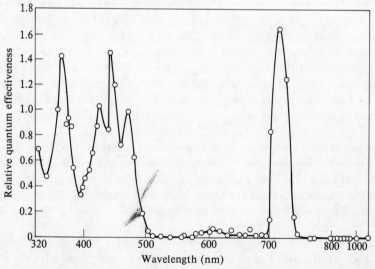

Fig. 3.9 An accurate action spectrum for the high-energy photoinhibition of lettuce hypocotyl growth (from Hartmann).[168]

These action spectra can tell us only one useful fact – that the two spectral regions of generally maximum activity are the blue and the red/far-red wavelengths. It is perhaps possible to speculate from the lack of a consistent relationship between the activities in these two regions, that the blue activity, and the red/far-red activity, are manifestations of two separate and distinct pigment systems, but these results are not conclusive on this point. As far as identification of the responsible photoreceptors is concerned, these action spectra are of little help, and it would be perfectly reasonable to conclude that each species utilizes a different photoreceptor, a situation which seems most unlikely.

The most accurate action spectrum yet obtained for a high-energy response is that for the inhibition of hypocotyl elongation in lettuce seedlings, constructed by Hartmann[168] at the University of Freiburg. As shown in Fig. 3.9, there is a very sharp action maximum which peaks at 717nm, and a complex series of peaks in the blue and the near ultraviolet regions. This elegantly constructed action spectrum is, by its very nature, indicative that the blue and far-red activities are mediated by different photochemical systems, since the pattern of peaks in the blue and ultraviolet is strongly reminiscent of a carotenoid or flavin photoreceptor, neither of which have any absorption in the red or far-red spectral regions.

It is instructive to compare this action spectrum with other carefully constructed action spectra for blue-light dependent photoresponses. A visual comparison with the action spectra for phototropism in *Avena* coleoptiles (Fig. 3.11) shows obvious similarities. Table 3.6 presents an analysis of the similarities of the action spectra for *Avena* coleoptile phototropism, for lettuce hypocotyl growth, and for carotenoid production in the fungus *Fusarium aquaeductum*. The similarities, especially between phototropism and the inhibition of hypocotyl elongation are striking, and can be taken as strong evidence that the photoreceptors are identical. The best evidence available for phototropism (see page 48) suggests that the photoreceptor is a carotenoid, although the possibility that it is a flavin or a flavoprotein has not been completely ruled out.

Table 3.6 Comparison of the peaks of maximum activity in the blue and ultraviolet regions for phototropism of *Avena* coleoptiles, the light inhibition of lettuce hypocotyl growth, and the light-mediated synthesis of carotenoids in *Fusarium aquaeductum*. The wavelength maxima (λ_{max}) and the relative effectiveness (R.E.) were calculated from published data (Curry[82]; Hartmann[168]; Rau[349]).

| | Peak numbers | | | | | | | |
| | I | | II | | III | | IV | |
	λ_{max}	R.E.	λ_{max}	R.E.	λ_{max}	R.E.	λ_{max}	R.E.
Phototropism	370	0·53	425	0·71	445	1·0	475	0·82
Hypocotyl growth	362	0·96	425	0·72	446	1·0	474	0·68
Carotenoid synthesis	377	0·74	438	0·76	451	1·0	477	0·92

The action spectrum evidence therefore indicates that two separate photosystems are involved in the high-energy reaction, one operating in the blue regions of the spectrum and involving a carotenoid or flavin photoreceptor, and the other operating in the red and far-red regions of the spectrum. We can derive no information on the identity of the photoreceptor for the longer wavelength processes from a consideration of the details of the action spectrum, and this problem will be taken up later.

3.4   Physiological evidence for two photoreceptors

Further evidence supporting the hypothesis that the blue and far-red activities are mediated through separate photochemical systems comes from experiments which show physiological differences between the responses to blue light, and to red or far-red light. The most conclusive evidence, in this case, comes from the interaction of the blue and far-red systems in the synthesis of anthocyanin and other phenolic compounds. Synthesis of anthocyanin in the turnip hypocotyl, for example, is dependent on irradiation of both the hypocotyl and the cotyledons; apparently a precursor is produced in the cotyledon which must be translocated to the hypocotyl before anthocyanin synthesis can take place there. Although continuous blue or far-red light causes pigment synthesis throughout the seedling, far-red light has a relatively greater effect in the cotyledon, while the two treatments have similar quantitative effects in the hypocotyl.[158]

Engelsma and Meijer[106] have made observations of a similar nature with gherkin seedlings, which synthesize hydroxycinnamic acids and flavonoids (phenolic secondary products), in response to continuous irradiation with blue or far-red light. They observed that blue light would cause an appreciable stimulation of phenol synthesis in excised hypocotyls, whereas far-red light had no effect unless the hypocotyls were attached to the cotyledons. Here again there is evidence that for far-red light to be active, translocation of an intermediate, or cofactor, from the cotyledons to the hypocotyl must be possible. It was also observed that the lag phase between the onset of irradiation and measurable inhibition of hypocotyl elongation was much shorter with blue than with far-red light[283] (see also Fig. 7.3). This effect is also detectable with lettuce hypocotyls and can be seen in the data of Fig. 3.5.

In turnip seedlings, when a 6 hour period of blue light was given before 42 hours of far-red, the total amount of anthocyanin produced was much smaller than if the 42 hours of far-red light were given without prior blue light treatment. When the experiment was repeated with far-red light preceding blue light there was no inhibition at all.[156] This inhibition of synthesis in a subsequent far-red irradiation period is also seen with pre-irradiation with red light. In this case marked inhibition is observed when the red irradiation precedes a prolonged exposure to far-red light, but very little effect is seen when the red irradiation precedes a prolonged exposure to blue light.[156] These results can only be explained on the basis that irradiation with blue and far-red light operates through different biochemical mechanisms. The simplest, and most likely, way in which this could happen would be if two different photoreceptors were being utilized. These experiments also provide useful information on the function of the photoreceptor for the far-red regions as is discussed in chapter 5.

We may therefore with a considerable degree of safety, conclude that two high-energy reactions exist operating through different photochemical molecules, and different biochemical control mechanisms. The most likely contender for the photoreceptor, for the blue action, is a carotenoid or a flavin.

It is useful to consider the potential adaptive value of the two photoreactive systems. As is discussed in chapter 2, the spectral regions of maximum incident solar energy at the earth's surface range from the blue region when the sun is overhead, to the red and far-red regions when the sun is near the horizon. Thus, the existence of high-energy, prolonged irradiation, photomorphogenic reaction systems with action maxima in these respective wavelength regions enables the photoregulation of development to take place under all conditions of latitude, and time of day.

3.5   The photoreceptor for the far-red action maxima

The action spectra shown in Figs. 3.7, 3.8 and 3.9 exhibit various maxima ranging from 660 nm to 730 nm. Since these peaks are in the red and far-red regions of the spectrum, an obvious candidate for the putative single photoreceptor is phytochrome. It is difficult, however, to see how phytochrome could be responsible for reactions with such different action maxima. If all the reactions had maxima at either 660 nm or 730 nm, it would be reasonable to propose that either Pr or Pfr was the absorbing molecule, but the action maxima situated between the absorption maxima of Pr and Pfr cannot be explained on this basis.

The first pointer to a deeper understanding of this problem came from a suggestion by Hendricks and Borthwick[181] in 1959 that high-energy reactions with action maxima in the red and far-red spectral regions could be mediated by the simultaneous absorption of light by the two forms of phytochrome, and their consequent simultaneous excitation. This concept was extended and elaborated by Butler and co-workers[64] in 1963 when they showed that Pfr was very much more unstable than Pr, and that the irreversible destruction or 'decay', of phytochrome, a process which is described in detail in chapter 5, occurred only with the Pfr form of the pigment. It was also discovered that Pr and Pfr do not only absorb light of wavelengths 660 nm and 730 nm respectively. These wavelengths are only the points of maximum absorption for each form of the pigment and in reality both Pr and Pfr have rather broad bands of absorption. This topic is covered in detail in chapter 4 and it is only necessary at this stage to point out that Pr and Pfr have overlapping absorption spectra. In fact, Pfr has considerable absorption at 660 nm, the wavelength of maximum absorption by Pr, and in turn Pr has slight, but significant absorption at 730 nm, the wavelength of maximum absorption by Pfr (see Fig. 4.9). Under continuous irradiation at wavelengths which are absorbed by both forms of the pigment, a dynamic equilibrium is rapidly set up in which the rate of conversion of Pr to Pfr is exactly balanced by the rate of conversion of Pfr to Pr. At equilibrium with a monochromatic light source a so-called 'photostationary state' is set up in which the relative proportions of Pr and Pfr are maintained constant at a value dependent on the wavelength. Continuous red light, therefore, which is mainly absorbed by Pr, will establish a photostationary state with a high proportion of Pfr, whereas continuous far-red, which is principally absorbed by Pfr will maintain only a very small proportion of Pfr.

If we then take into account the observations that Pfr is very unstable, whereas Pr is stable, we can see that under continuous red light, the rate of loss of total phytochrome is high, due to loss of the Pfr form, whereas under continuous far-red light, the rate of loss of total phytochrome is low, since only very small amounts of Pfr are available. If it is assumed that the high-energy reaction demands the maintenance of Pfr for a long time, but not necessarily at high concentrations, it can be seen that far-red light will be most effective. Furthermore, since all wavelengths between the absorption maxima of Pr and Pfr are absorbed by the two forms setting up characteristic photostationary states, it only remains to postulate that in different responses different proportions of Pfr/Ptotal produce maximal developmental response, to provide an explanation for the detailed differences in the action maxima.

These ideas in themselves, although attractive, do not constitute proof, and it was not until 1966 that Hartmann[167] reported direct experimental evidence for the role of phytochrome as the photoreceptor for the far-red high-energy reaction. In this most elegant work, the principle was to show that maximum photomorphogenic response was obtained with a specific proportion of Pfr/Ptotal, irrespective of the manner whereby this photostationary state was achieved.

The high-energy response investigated was, again, the photoinhibition of hypocotyl lengthening in lettuce seedlings. The action spectrum for this response (Fig. 3.9) shows a very sharp maximum at about 716 nm, a wavelength which establishes a photostationary state of Pfr/Ptotal of 0·03. Thus, it can be postulated that for continuous irradiation, maximum effect is observed when Pfr is maintained at about three per cent of the total phytochrome. It is possible to establish this photostationary state by irradiating the tissue with precisely balanced energies of two beams of light of wavelengths near the absorption maxima of Pr and Pfr respectively. In Hartmann's experiments the tissues were irradiated with a standard quantum flux density of 2400 picoEinsteins $cm^{-2} s^{-1}$ at 768 nm (which is only absorbed by Pfr and has no inhibiting effect alone), together with varying energies of 658 nm (which is absorbed principally by Pr and which also has no inhibitory effect alone).

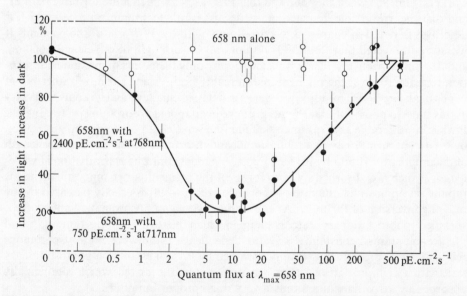

Fig. 3.10 The dual-wavelength experiment of Hartmann. Maximum inhibition with a mixture of 768 nm and 658 nm radiation occurs at precise energy levels in which the photostationary state is approximately 0·03 (reprinted with permission from Hartmann).[167]

In this way, the proportions of Pfr/Ptotal could be varied from 0·0 to the theoretical maximum of 0·8 (see chapter 4). The results of this experiment are given in Fig. 3.10 and they show clearly that as the 658 nm quantum flux density is increased with a standard energy of 768 nm, the biological response ranges from zero, through maximum inhibition (at about 10 pE $cm^{-2} s^{-1}$ of 658 nm) to zero, once more, at high irradiances. The photostationary state at the point of maximum inhibition was calculated and found to correspond exactly with that established with monochromatic light at the action maximum, i.e., Pfr/Ptotal is 0·03. This experiment can be carried out by mixing any two monochromatic wavelengths which will establish the whole range of possible photostationary states, and Hartmann has shown that whatever wavelengths are used, maximum inhibitory effect is always obtained at a Pfr/Ptotal of 0·03.

These results are extremely convincing evidence that phytochrome is the photoreceptor for the far-red action maximum of the high-energy reaction, and that the wavelength of maximum action simply represents that wavelength which sets up, and maintains, the most favourable photostationary state for the particular response. For the inhibition of lettuce hypocotyl extension, this critical photostationary state is 0·03 and levels of Pfr above or below three per cent of total P are very much less efficient, as is evidenced by the sharpness of the action maximum (Fig. 3.9). There is nothing magical about the 0·03 photostationary state, however, since, even in the same species, there is a high-energy reaction which requires a different value of Pfr/Ptotal for maximum effect. This response is the inhibition by continuous high-energy light of the germination of lettuce seeds already potentiated to germinate by brief treatment with red light. By using the methods outlined above, Hartmann has shown that the optimum photostationary state for this high-energy reaction is 0·1, i.e., ten per cent of total phytochrome must be in the Pfr form.[167]

A word of caution is necessary here, however. Schneider and Stimson[378a] in 1971 pointed out that although the high-energy reaction is normally found in etiolated tissues, nevertheless there may be sufficient chlorophyll present, especially after an hour or so of irradiation, for the light reactions of photosynthesis to be operating. They referred to earlier findings of Arnon et al.,[5a] that non-cyclic photophosphorylation had an action maximum around 660 nm, whereas that for cyclic photophosphorylation was between 700 and 720 nm. In addition, antimycin A and 2,4-dinitrophenol, which are known to inhibit cyclic photophosphorylation, also inhibited anthocyanin synthesis, induced by far-red light, in etiolated turnip hypocotyls.[378a] On the other hand, 3-(3,4-dichlorophenyl)-1,1-dimethylurea (DCMU) and o-phenanthroline, which inhibit non-cyclic and stimulate cyclic photophosphorylation, actually promote anthocyanin synthesis. Thus, Schneider and Stimson[378a] suggest that during the far-red light treatment, sufficient chlorophyll is produced to allow cyclic photophosphorylation to proceed, and that this light reaction is responsible for the high-energy effect. In another high-energy light reaction, the opening of fully green Mimosa pudica leaflets, far-red light drives the response, even if simultaneous irradiation is given which varies the Pfr/Ptotal proportion over a wide range.[122a]

It is not yet possible therefore to conclude definitively whether phytochrome or chlorophyll is the photoreceptor for the far-red band of action in the high-energy reaction. The balance of the evidence, however, is currently in favour of phytochrome as much of the evidence for chlorophyll is dependent on inhibitor effects which can never be wholly reliable due to the possibility of unknown side-effects. Furthermore, the precise relationship between Pfr/Ptotal and response observed by Hartmann, together with the extremely sharp action maxima, is hard to explain on the basis of light absorption by chlorophyll. A possible compromise, suggested by Schneider and Stimson,[378a] is that both chlorophyll a and phytochrome are involved; in this case, however, chlorophyll may merely be operating via energy transduction, rather than through a true photomorphogenic mechanism.

3.6   The phototropic response system

Phototropism is the phenomenon in which a directional light stimulus elicits a directional growth response in a plant organ. Thus, phototropism is distinguished from photomorphogenesis, in which the direction of the developmental response is not related to the direction of the light stimulus, nor is it dependent on the light stimulus being of a directional nature. This definition also separates phototropism from photonasty, in which movement of whole plant organs, in response, usually, to a non-directional light stimulus occurs by virtue of turgor changes in specialized pulvinal cells, and does not involve growth changes. Photonastic movements, although of importance in nature in determining the orientation of leaves, flowers, and petals in relation to the incident light, do not directly affect the development of the plant, and thus are outside the immediate scope of this book.

Phototropism is of great importance in determining the direction of growth of plant organs in natural environments. The responses are, by definition, restricted to those parts of the plant with a potential for growth and can be found in the growing stems, leaves, and sometimes roots, of higher plants, in the chloronemata of ferns, the sporophores of mosses, and the sporangiophores of certain fungi. Most of the analytical work on this phenomenon has been carried out with grass coleoptiles or fungal sporangiophores, and although much has been achieved, it is not certain how far the concepts derived from this exact, but restricted, experimentation can be extended to cover the behaviour of other organs under natural conditions.

(a) Development of the phototropism concept

The recognition that plant organs bend in response to unilateral illumination must surely have occurred very early in the history of botany. The biggest single step forward, however, was probably made by Charles Darwin[83] in work beautifully described in his *Power of Movement in Plants*, published in 1880 and highly recommended for student study. In this work he clearly demonstrated the existence of substances emanating from the tip of a coleoptile which regulate growth in the basal regions of the coleoptile. Such substances, later known as auxins, were not isolated and characterized until nearly 50 years later. Darwin, utilizing daylight, paraffin lamps, and wax candles as light sources, showed by ingenious shading methods that lateral illumination of the apical 0·25 inches of the coleoptile resulted in curvature, but that the growth responsible for the curvature developed in the shaded zone below the tip. Only slight curvature was observed when the apical 0·25 inches was shielded, and the basal parts illuminated. These experiments suggested that the apex is the region of maximum photosensitivity, and upon unilateral illumination transmits some influence to the lower parts of the coleoptile bringing about an asymmetric change in growth rates.

Darwin's hypothesis was to some extent challenged in 1915 by Blaauw[22] who analysed the so-called light-growth reaction in *Avena* coleoptiles and *Phycomyces* sporangiophores, in which a change in overall growth rate is caused by a change in irradiance. Blaauw developed the hypothesis that a gradient of irradiance across the coleoptile would lead to a similar, but inverse, gradient of growth rate as a result of the light-growth reactions. This hypothesis does not take into account Darwin's localization of the region of maximum phototropic sensitivity in the coleoptile apex. It must be stressed, however, that modern experiments have shown that phototropic sensitivity is not absolutely restricted to the apex, and that short-term stimulation of the lower regions does lead to small curvatures. Continuous illumination of these regions causes much greater curvatures. Thus, the two concepts are not completely incompatible.

The next major development in the understanding of the mechanism of phototropism awaited the experimental demonstration of the auxin hormones. In 1927 and 1928, Cholodny[70] and Went[454] independently postulated that unilateral illumination of a coleoptile established a lateral component in the predominantly basipetal transport of auxin from its site of synthesis, the apex, resulting in an accumulation of auxin on the darker side of the organ, and a reduction on the illuminated side. The asymmetric distribution of auxin was thus held to be responsible for the changes in growth rate on the two sides of the coleoptile.

A possible alternative to lateral transport as the mechanism for achieving asymmetric distribution of auxin was proposed by Galston and Baker[138] who discovered that the auxin indole acetic acid (IAA) was destroyed *in vitro* when illuminated in the presence of riboflavin, a possible photoreceptor for phototropism. However, very little *in vivo* evidence supporting the photolysis of auxin on the illuminated side has been obtained and, in fact, Briggs[56] has shown that light does not decrease the total yield of auxin diffusing into agar blocks from the bases of isolated coleoptiles. Direct observation of lateral transport of radioactively labelled IAA[85, 326] has since been obtained proving the validity of the Cholodny–Went hypothesis, which therefore remains the most plausible explanation of phototropism in coleoptiles and which has survived the years with only slight modifications. The problems facing the present-day investigator of phototropism are, on the one hand, the elucidation of the nature of the photoreceptor and its mode of action in regulating auxin distribution and, on the other hand, to determine whether or not the concepts of phototropism in the coleoptile have relevance for plants growing in the field.

3.7 The phototropism photoreceptor

(a) Action spectrum studies

As has been stated earlier, the most powerful method for the investigation of the nature of the photoreceptor for any light-mediated response is to attempt to match the action spectrum of the response with the absorption spectra of putative photoreceptor pigments. This approach has been applied extensively to phototropism but has not as yet yielded an unequivocal answer.

A very carefully constructed action spectrum for the first positive photoresponse (see below) in *Avena* coleoptiles is given in Fig. 3.11. There is no activity beyond 500 nm, i.e., in the green, yellow, and red regions of the spectrum, while there are two pronounced peaks of activity in the blue, at 445 nm and 475 nm, and a shoulder at about 425 nm. There is a further peak in the near ultraviolet at 370 nm.

The restriction of activity to the blue regions of the spectrum suggests that the pigment is a yellow compound, since only a yellow substance would absorb blue light but not red light. The finer detail of the action spectrum has enabled the choice of photoreceptor to be restricted to one of a few specific substances known to be present

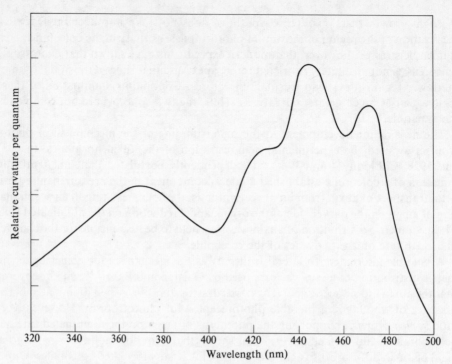

Fig. 3.11 The action spectrum of phototropism in *Zea mays* coleoptiles (after Curry).[82]

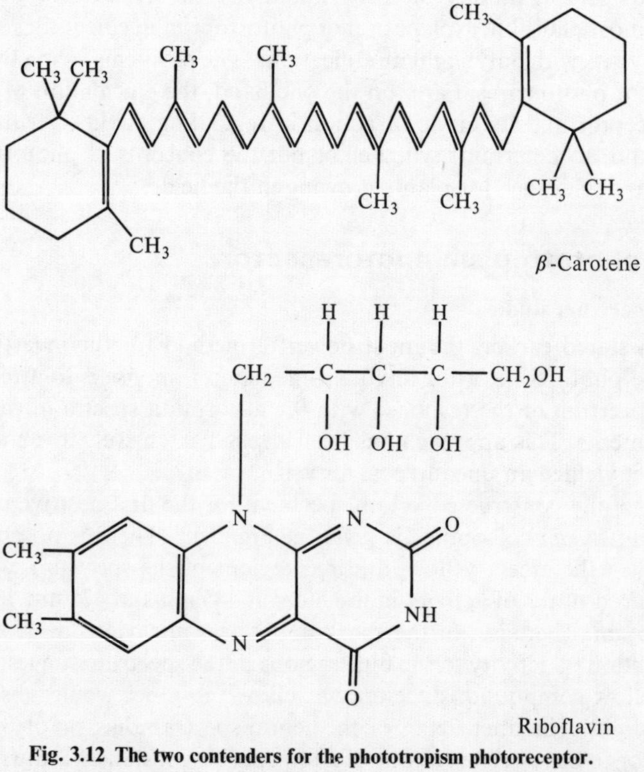

β-Carotene

Riboflavin

Fig. 3.12 The two contenders for the phototropism photoreceptor.

in plants, the most popular candidates being a carotenoid, or a flavin, (Fig. 3.12). Carotenoids tend to have characteristic multiple absorption peaks in the blue region which match reasonably well with the action spectrum in that region (Fig. 3.13). The known carotenoids, however, do not have a peak of absorption at or near 370 nm, although the existence of such a compound has been postulated on chemically acceptable grounds. Flavins do not have a similarly complex pattern of absorption in the blue region, although riboflavin has a strong peak at 370 nm which matches well with that region of the action spectrum (Fig. 3.13).

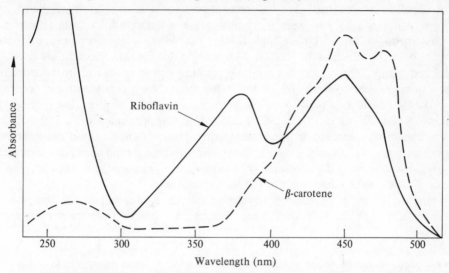

Fig. 3.13 Absorption spectra of riboflavin (solid line) and β-carotene (dotted line) (from Curry).[82]

The action spectrum evidence thus does not provide us with a definitive answer concerning the photoreceptor. The ambiguous nature of the action spectra may be due to a considerable number of causes. The active pigment may exist *in vivo* as a complex with another component in which the absorption spectrum is drastically changed. The existing extraction methods may be insufficiently 'gentle' for the preparation of the complex without dissociation. Another possibility is that masking by other absorbing substances may change the apparent action spectrum. A final possibility is that a carotenoid and a flavin pigment may cooperate in light absorption for phototropism, thus providing a composite action spectrum. These questions cannot be resolved at the present time and further progress must await refinements in techniques.

(b) Distribution of sensitivity and pigments

Darwin[83] observed that the region of greatest sensitivity in the oat coleoptile was the tip region, i.e., the apical 0·25 inches. However, in more exact experiments with the *Phalaris* coleoptile, he found that the apical 0·05 inches was virtually insensitive to light, while illumination of the region between 0·15 and 0·2 inches from the tips was extremely effective in causing bending. More modern evidence corroborates these findings, and appears to indicate that the most sensitive region of the oat coleoptile is approximately 100 μm from the extreme tip, with a decreasing gradient in sensitivity down the organ.[82, 257]

The putative photoreceptor would be expected to reflect this gradient in photosensitivity with a similar gradient in concentration. Both flavins[5] and carotenoids[405] have been found to be concentrated in the apical regions and to exhibit a decreasing concentration gradient below. In the case of the carotenoids, a high concentration has been found just below the tip with none detectable in the extreme apex. Thus the distribution of the carotenoids, at least, appears to reflect that of tropic photosensitivity.

Recent work of Curry[82] has provided stronger evidence that pigment concentration parallels photosensitivity. In preparations of squash coleoptiles, bands of yellow coloured cells were seen surrounding the vascular bundles, and stretching almost up to the apex. Under high power, the yellow pigments appeared to be localized in the plastids, organelles which normally contain carotenoids even in etiolated plants. The pigment-containing plastids were not present in the extreme apex, but in the region from 100 μm to 500 μm from the tip, considerable concentrations of the coloured plastids occurred in the cells near the bundles. The number of coloured plastids per cell decreased rapidly below approximately 2 mm from the apex. The *in vivo* absorption spectrum of these plastids showed that considerable concentrations of carotenoids were present there, and the strong correlation between plastid distribution and photosensitivity is evidence to suggest that the carotenoid in the plastids is the photoreceptor for phototropism.

Thus, the available evidence is slightly in favour of a carotenoid as the photoreceptor, although the anomalies in the absorption and action spectra remain to be explained.

An interesting correlation may be drawn at this point with the phenomenon of geotropism, in which plant organs respond to a directional gravitational stimulus by a directional growth response. The available evidence implicates the so-called statoliths (plastids containing large amounts of starch as starch grains) as the receptors of the gravitational stimulus, and suggests that the movement of these organelles across the cell is important in the cellular mechanism of geotropism.[393] The evidence that the photoreceptor for phototropism may also be localized in plastids suggests that plastids may be of general importance in the mediation of the asymmetric growth of the cells.

3.8 Dose-response relationships

The relationship between amount of light energy given and the angle of curvature produced is exceedingly complex. A typical dose-response curve for oat coleoptiles is shown in Fig. 3.14. It can be seen that this curve consists of three major regions. At low energies, the curvature is positive, i.e., toward the light source, and the amount of curvature is proportional to the logarithm of the energy applied. As the total energy is increased, the amount of curvature decreases until, over a short range of energies, small curvatures away from the light source are observed. Both of these responses to relatively low total energies are dependent on the light being given over a short period (i.e., less than 2–3 minutes). At higher energy levels, no curvature at all is observed, unless the time of irradiation exceeds 3–4 minutes, whereupon marked curvatures towards the light, along the whole of the coleoptile are seen.

The three regions of the dose-response curve are known as 'first-positive', 'first-negative', and 'second-positive'. It should be pointed out here that almost all the analytical work on the nature of the photoreceptor and the mechanism of growth curvatures has been restricted to the 'first-positive' response. This is to an extent unfortunate, since the phototropic responses of plants growing under natural conditions are normally due to continuous illumination with high irradiances and thus fall into the 'second-positive' category.

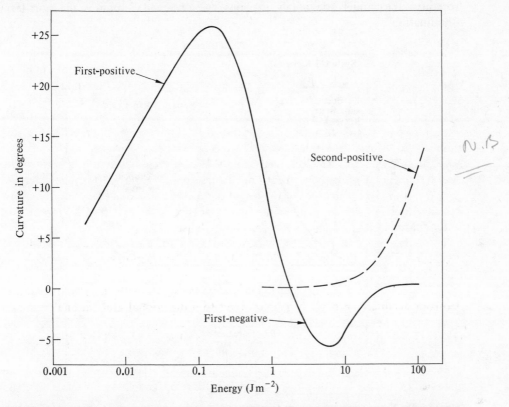

Fig. 3.14 Dose-response relationships for coleoptile phototropism. The second-positive response (dotted line) requires radiation over a longer period than the first-positive and the first-negative responses (modified from Curry).[82]

It has proved extremely difficult to reconcile the complex dose-response relationship with a simple hypothesis for photochemical action in phototropism. The first-positive and first-negative responses obey the Bunsen–Roscoe reciprocity law,[49] which states that when only one photoreceptor is operating, the photochemical effect of light remains the same if the product of light intensity and duration of irradiation remains the same (chapter 2). Thus, it seems likely that only one photoreceptor is involved in these low-energy responses. The second-positive response, however, does not obey the reciprocity law; in fact, long exposures at low irradiances can result in marked second-positive type curvatures even though the total energy falls into the first-negative category.[49] Thus, for the second-positive curvatures, some process depending on the time of exposure, and not on total energy, must be postulated.

The most recent attempt to produce a model scheme to account for the oscillations in the dose-response curve is that of Briggs,[49] in which each of the three responses is considered to be mediated by a separate photoreceptor molecule. The model is schematically represented in Fig. 3.15. The photoreceptors for Systems 1 and 2, which correspond essentially to the first-positive and first-negative responses, are thought to be transformed from inactive to active forms by low-energy light, and subsequently to further inactive forms by higher energy light. Assuming the photoreceptor for System 1 to be slightly more sensitive to light than the System 2 photoreceptor, the model adequately explains the observed responses to short-term illumination.

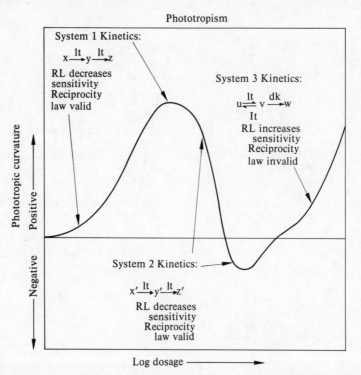

Fig. 3.15 The Briggs scheme to account for the complicated dose-response curves of phototropism (from Briggs).[49]

The photoreceptor for System 3 is again photoconverted to an active form, this time by much higher energies of light, but this active form is thought to be either photochemically reconverted to the original inactive, but photosensitive form, or thermochemically converted to a new inactive, non-photosensitive form.

This complex model adequately explains the complex situation, but it would seem unlikely that plants have evolved three completely separate mechanisms for reactions to different energies of unilateral illumination. It seems more reasonable to assume, as Curry[82] has pointed out, that the dose-response relationship observed is a consequence of the experimental methods chosen, especially in relation to the somewhat artificial techniques of very brief illumination treatments.

3.9   Interaction with phytochrome

Since the first action spectrum studies had shown that wavelengths longer than *ca.* 500 nm elicited no phototropic responses, it was initially considered safe to use red light sources for the manipulation of the experimental material. Such dim red safe-lights were used in virtually all early experiments on the dose-response relationships, action spectra, and distribution of photosensitivity in phototropism. In 1957, Curry[81] discovered that a pretreatment with non-directional red light markedly affected the sensitivity of the coleoptile to a subsequent directional, blue light stimulus. The effect manifests itself as a decrease in the sensitivity of the first-positive response (Fig. 3.16), and an increase in the sensitivity of the second-positive response. The first-negative response may disappear altogether due to the overlapping of the two positive responses.

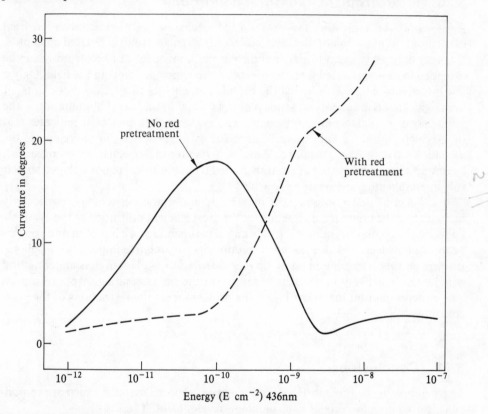

Fig. 3.16 The effect of a preirradiation with red light on phototropic sensitivity of maize coleoptiles – the first-positive curvature is moved to the right along the energy axis, i.e., the phototropic sensitivity is reduced (after Chon and Briggs).[71]

The precise timing of the pretreatment with red light does not appear to be important, and any red light exposure given between germination and the time of experimentation is sufficient to modify the sensitivity of the organ. Red light given after unilateral blue light, however, has no effect on the subsequent phototropic response. The effect of red light pretreatment has been intensively studied by Briggs[50, 71] who has shown that the effects are reversible by far-red light, thus invoking the involvement of phytochrome. The precise details of the phytochrome involvement have led to serious questions concerning some other previously accepted ideas on phytochrome action, as is described in more detail in chapter 5.

The mechanism of phytochrome action in controlling the phototropic sensitivity of the coleoptile is as obscure as is its mechanism in other responses. It is significant, however, that pretreatment with red light also affects the geotropic sensitivity of coleoptiles, in a manner which again suggests phytochrome mediation,[458] implying that phytochrome is involved in the mechanism for achieving asymmetric growth, rather than in the mechanism for the perception of the directional stimulus.

3.10   Phototropism in natural conditions

As stated above, almost all the work on the mechanism of phototropism has been carried out on grass coleoptiles, and much of it restricted to the etiolated coleoptile. It is very difficult to say how much of the current knowledge of phototropism can be extended in an unmodified form to cover the behaviour of plants in the field. Only a few investigators have considered the problem of whether dicotyledonous plants, or even light-grown monocotyledonous plants, respond to unilateral illumination in the same way as etiolated monocotyledons. The evidence that does exist indicates that light-grown plants are only sensitive to relatively high energies, in the region of the second-positive response for oat coleoptiles.[47] As far as the mechanism is concerned, however, it seems likely that auxin transported from the apical regions plays a similar role in establishing asymmetric growth.[86, 474]

It is believed that phototropism may be of importance in ecology, particularly in regulating the orientation of plants for the most efficient utilization of the available light energy for photosynthesis. Hardly any experimental work has been done on this problem, however, and it is not even certain what the relative importance of phototropism and geotropism is in regulating the orientation of plant organs under natural conditions. It is likely that further progress in the understanding of phototropism will be dependent on the liberation of the problem from the restrictions of the grass coleoptile.

3.11   Conclusions

There appear to be three distinct photoreactive response systems which are responsible for the direct effects of light on plant development. These are:

1. the low-energy phytochrome system;
2. the high-energy phytochrome system; and
3. the high-energy blue system.

In addition the phototropic response system mediates the directional responses of plants to directional light stimuli.

The simplest conclusion is that only two photoreceptors are necessary to explain the observed data, one photoreceptor being phytochrome, the other being an, as yet, unidentified blue-absorbing substance, which is likely to be either a carotenoid or a flavin. It must be stressed that these are the simplest conclusions, and it is only too likely that other photoreactive response systems will be defined in the future. It is certainly true, for example, that fungi exhibit certain photomorphogenic and phototropic responses that do not fit in any of the categories described here.

The rest of this book is concerned with an analysis of phytochrome and its regulation of plant development. This concentration on phytochrome should not be taken to indicate that phytochrome is more important than the blue-absorbing photoreceptor; merely that at the present time, we know very little about the latter and considerably more about phytochrome.

4

The chemistry of phytochrome

The realization of the importance of phytochrome as a regulator of plant development in the late fifties led to intensive efforts by several research groups to isolate,
purify, and chemically characterize the pigment. The logic behind this drive was the
hope that knowledge of the chemical structure and properties of phytochrome would
provide clues to the nature of the biochemical mechanism of phytochrome action
within the plant. Whether or not this hope will be realized remains to be seen, but it
is already clear that phytochrome is a most unusual molecule of great intrinsic
interest to physical and biological chemists.

4.1 Detection and measurement of phytochrome

The first demonstration of the existence of phytochrome in plant tissues, as described
in the preceding chapter, was carried out with the aid of a single-beam recording
spectrophotometer specially designed for dense, light-scattering samples. With this
instrument, Butler et al.[65] were able to show that actinic red and far-red light treatments produced photoreversible changes in the absorption spectrum of living tissues.
When the tissues were irradiated with red light to photoconvert Pr to Pfr, a decrease
in absorption was seen at about 660 nm associated with an increase in absorption at
730 nm, as had been predicted from the earlier physiological observations. Irradiation
with far-red light substantially reversed these spectral changes (see Fig. 3.2). It was
therefore clear that phytochrome could be detected, and therefore presumably
measured, in living tissues by virtue of these red/far-red photoreversible absorbancy
changes. Butler et al.[65] also showed that similar photoreversible absorbancy changes
were detectable in buffered aqueous extracts of etiolated plant tissues.

With the exception of recent, rather complicated, immunological procedures, the
measurement of a red/far-red photoreversible absorbancy change is still the only
assay for phytochrome either in vivo, or in vitro. It is not known what metabolic
reactions, if any, are entered into by phytochrome, and thus it is not yet possible to
devise a biochemical or enzymatic assay for phytochrome. Consequently, the majority
of the information we have on the physiological and chemical properties of phytochrome is based on the spectrophotometric measurement of the reversible absorbancy
changes. In contrast with an enzymatic assay, which has its own built-in amplification
device, due to the capacity of one enzyme molecule to catalyse the conversion of many
thousands of substrate molecules, the photometric assay is relatively insensitive. As
will be seen, this raises considerable problems during the extraction and purification
of phytochrome, since it is always necessary to work with large amounts of plant

material. An understanding of the techniques used to measure the spectral changes is therefore necessary.

The spectrophotometer used initially by Butler et al.[65] was capable of recording full spectra, but suffered from a grave disadvantage due to its being only a single-beam instrument. This meant that the signal produced was not only a function of the optical properties of the tissue sample, but also included the system response of the instrument itself. Consequently, other instruments have been devised which overcome this problem.

The most favoured approach has been to use a dual-wavelength difference photometer which measures the absorbance at 660 nm and at 730 nm in rapid alternation, and electronically computes the difference in absorbance at the two wavelengths. Figure 4.1 is a diagrammatic representation of the first such instrument used in phytochrome work.[65]

The sample cuvette C is held in close juxtaposition to the photomultiplier tube P, thus allowing dense, light-scattering samples to be measured. Two monochromatic beams are isolated by interference filters F_1 and f_1 and F_2 and f_2 respectively, and directed through the sample cuvette by the lenses L_1 and L_2 and the mirror systems M_1 and m_1, and M_2 and m_2. It is necessary for these beams to be balanced with neutral density filters (not shown). The rotating chopper blades Ch_1 and Ch_2 cause the sample to be irradiated alternately with the two beams via a diffusing disk d. The chopper blades are synchronized to the phototube so that the measuring circuit automatically records the difference in absorbance at the two wavelengths:

$$\text{(i.e., } \Delta A = A_{660} - A_{730}).$$

The principle of phytochrome detection and measurement using a dual-wavelength difference photometer is based on the fact that Pr has a high specific absorbance at 660 nm and Pfr has a high specific absorbance at 730 nm. Thus, a tissue or solution sample which contains phytochrome solely in the Pr form will have a higher absorbance at 660 nm, and a lower absorbance at 730 nm, than a sample containing largely Pfr. Upon photoconversion with red light, the sample will exhibit a spectral change leading to a decreased absorbance at 660 nm and an increased absorbance at 730 nm (as in Fig. 3.2.). This change in absorbance is directly related to the phytochrome content of the sample, and thus can be used to estimate the relative amounts of total phytochrome in the tissue. Furthermore, it is possible to measure the relative amounts of Pr and Pfr in the tissue as prepared. The simplest procedure used is as follows:

1. Prepare the sample and place in the instrument (in a darkroom).
2. Irradiate the sample with an actinic source of red light to photoconvert Pr to Pfr.
3. Determine ΔA (i.e., $A_{660} - A_{730}$) . . . ΔA_r.
4. Irradiate the sample with an actinic source of far-red light to photoconvert Pfr to Pr.
5. Determine ΔA . . . ΔA_{fr}.

The relative amount of total phytochrome is now related to the difference between the ΔA readings after red and far-red irradiation:

Thus, $$\Delta(\Delta A) = \Delta A_{fr} - \Delta A_r$$

This assay measures the sum of the photoreversible Pr and Pfr absorbances, and thus $\Delta(\Delta A)$ is directly related to the phytochrome content. In a great deal of the earlier

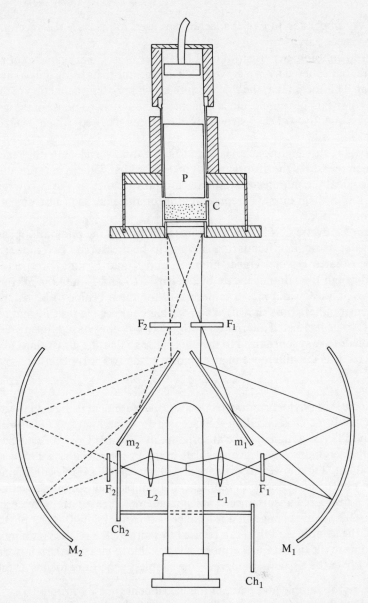

Fig. 4.1 Diagram of the first dual-wavelength photometer used for the estimation of relative phytochrome levels in etiolated plant tissues (from Butler *et al.*).[65]

literature, absorbance was referred to as optical density, and thus the notation ΔOD and $\Delta(\Delta OD)$ was used in place of ΔA and $\Delta(\Delta A)$. A modification of this technique is to use 730 nm as the single measuring wavelength and 800 nm as a reference wavelength (see below).

The above procedure can best be understood by considering the actual changes in absorption spectrum of a sample of tissue. In Fig. 4.2 the upper curve represents the absorption spectrum of maize shoots (*cf.* Fig. 3.2). An average etiolated tissue sample will have an overall A of 3 to 4, but only a small fraction of this (about 5×10^{-2} A) is due to the phytochrome present. In Fig. 4.2 the major peak at *ca*. 670 nm is due to chlorophyll, and no phytochrome peak as such can be discerned. Difference spectra, however, show the overall spectral changes obtained upon photoconversion of phytochrome (see Fig. 3.2). The actual ΔA measurements with either 660/730 nm or 730/800 nm measuring systems are shown on the right-hand side of Fig. 4.2.

In both cases a relative measure of total phytochrome is given by:

$$\text{Ptotal} = k\ \Delta(\Delta A) = k(\Delta A_{fr} - \Delta A_r)$$

where k is an unknown constant of proportionality.

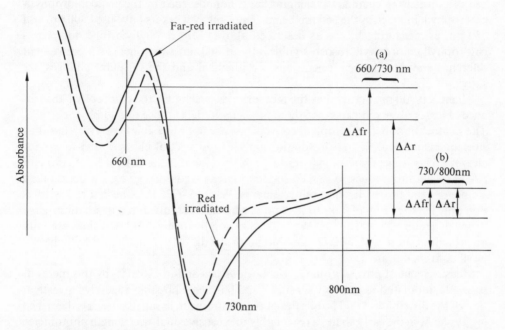

Fig. 4.2 The procedure for measuring relative phytochrome levels using a dual-wavelength difference photometer:
(a) using 660 nm and 730 nm measuring filters,
(b) using 730 nm and 800 nm measuring filters.
———— = absorption spectrum of red-irradiated tissue
– – – – = absorption spectrum of far-red irradiated tissue
ΔA_r = difference in absorbance at the two wavelengths for red-irradiated tissue
ΔA_{fr} = difference in absorbance at the two wavelengths for far-red irradiated tissue
(see text for procedural details).

This calculation is, of course, based on the assumption that red light converts *all* Pr to Pfr and that far-red converts *all* Pfr to Pr. In fact, as is discussed in detail on page 73, while far-red light is almost totally effective in converting Pfr to Pr, red light sets up an equilibrium between the two forms with approximately 81 per cent Pfr and 19 per cent Pr. For this reason the above calculation is normally modified as follows:

$$\text{Ptotal} = 1{\cdot}25\ k\ \Delta(\Delta A) = 1{\cdot}25\ k\ (\Delta A_{fr} - \Delta A_r)$$

If it is desired to determine the relative amounts of Pr and Pfr in a tissue sample it is general practice to reverse the sequence of irradiations given above so that actinic far-red light is given first. This converts any existing Pfr to Pr and allows direct estimation of Pfr; Pr is then determined by difference. This modified procedure was evolved to overcome, in part, one of the major difficulties associated with the photometric measurement of phytochrome in living tissues, namely the absorption of red light by protochlorophyll, which is thereupon converted to chlorophyll a. This photochemical conversion results in a change in the absorption spectrum of the tissue which is unrelated to phytochrome, and which thus interferes with phytochrome determinations. Measuring the Pfr content before the tissue is irradiated with red light partially overcomes this problem, but a much more satisfactory method is to use as a measuring beam a wavelength that is not absorbed by the protochlorophyll-chlorophyll, in place of the 660 nm beam. Thus, many workers have used 800 nm and 730 nm as measuring beams as described above; neither phytochrome nor protochlorophyll-chlorophyll absorb significantly at 800 nm which acts as a convenient reference wavelength. The operations are identical and the resultant changes are shown in Fig. 4.2.

When 660/730 nm are used as the measuring wavelengths, the chlorophyll absorbance changes observed can be of the same magnitude as those due to phytochrome. The protochlorophyll-chlorophyll conversions are not reversible, however, and thus subsequent red and far-red irradiation only cause spectral changes due to phytochrome. This is clearly seen in Table 4.1 which compares the ΔA (665/725 nm) and ΔA (725/800 nm) values, observed in *Amaranthus* seedlings, given a succession of actinic red and far-red irradiation treatments. With 665/725 nm, the first red irradiation only led to a drop of 4×10^{-3} A units whereas subsequent red irradiations gave drops of around 25×10^{-3} A. With 725/800 nm, the overall $\Delta(\Delta A)$ values are only about half of those with 665/725 nm, but no problems with protochlorophyll-chlorophyll conversions are seen.

Measurement of phytochrome levels, both *in vivo* and in extracts, by this method, is simple, rapid and reasonably sensitive, and has made possible a massive investigation of the distribution and properties of phytochrome in living plants, as is described in chapter 5. In the early sixties a reasonably inexpensive dual-wavelength photometer became commercially available.[21] This instrument, which is known as the Ratiospect, was made by the Agricultural Specialty Company of Beltsville, Maryland, and is shown in diagrammatic form in Fig. 4.3. The Ratiospect is normally modified to include a built-in irradiation system for the photoconversions. Although the Ratiospect has been greatly used in the last decade, and has been the source of much information, it does not have a particularly high sensitivity. In typical cases, the lowest level of detection using the Ratiospect is reached when the total $\Delta(\Delta A)$ of the sample being measured is about 1×10^{-3} A. This limitation has led to problems in the investigation of some of the *in vivo* properties of phytochrome.

Certain workers have sought to overcome this problem by constructing dual-wavelength instruments of much greater sensitivity. Figure 4.4 shows a diagram of one such instrument, designed and constructed by Dr C. J. P. Spruit[406] at Wageningen in the Netherlands, which has a sensitivity as low as 2×10^{-5} A. With this instrument, Spruit has been able to make very important and far-reaching discoveries relating to phytochrome in dry and imbibed seeds (see chapters 5 and 7). Spruit[407] has also constructed a dual-wavelength quasi-continuous recording spectrophotometer with which he has been able to investigate the steady-state levels of phytochrome intermediates at normal and very low temperatures (see page 95). Investigations of the spectral properties of purified phytochrome can be carried out with a range of commercially available spectrophotometers.

Table 4.1 Phytochrome measurements in dark-grown seedlings of *Amaranthus caudatus*. Two hundred 3-day-old seedlings (750 mg fresh weight) were packed to a depth of 10 mm in a cuvette of 12 mm diameter (from Frankland[123]).

Measuring wavelengths nm	D	FR	R	FR	R	FR	R	FR	R	Average $\Delta(\Delta A) \times 10^3$ A
				$\Delta A \times 10^3$ A						
665/725	19	20	16	47	22	51	25	54	26	26·3
725/800	35	35	55	39	56	40	57	41	—	16·0

The estimation of phytochrome, *in vivo*, using a dual-wavelength photometer of any kind, only provides a relative value. It is not possible from $\Delta(\Delta A)$ values to determine the absolute amount of phytochrome present in the tissue. Furthermore, the mode of sample preparation is very critical and it is not possible to make very meaningful comparisons between the $\Delta(\Delta A)$ values of separate samples, unless great care is taken to ensure uniformity of preparation. Equal weight tissue samples are usually

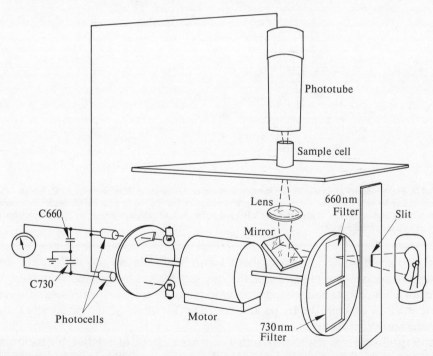

Fig. 4.3 Diagram of the 'Ratiospect', the dual-wavelength difference photometer used for phytochrome measurements. A synchronous motor rotates a filter wheel holding a 730 nm interference filter, and either a 660 nm or a 800 nm interference filter such that the tissue is alternately irradiated with the two light beams. The difference in light detected by the photomultiplier is computed electronically into a $\Delta(A)$ reading. The small lamps and photocells are used to synchronize the electronic system with the measuring beams. This instrument is able to detect $\Delta(\Delta A)$ values as low as 10^{-3} in samples of 2 or 3 A (from Birth).[21]

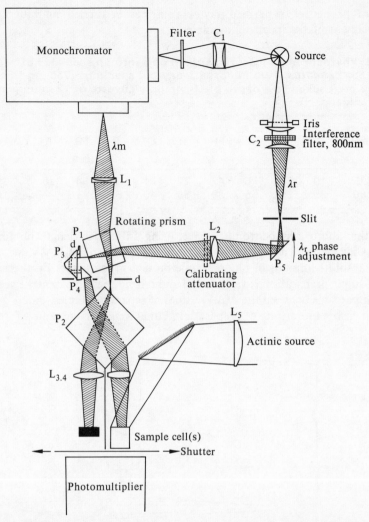

Fig. 4.4 A highly sensitive, dual-purpose spectrophotometer designed by Professor C. J. P. Spruit. This instrument can be used either as a dual-wavelength photometer (as shown) or as a double-beam instrument, in which case a second cuvette is placed in the left-hand beam. $\Delta(\Delta A)$ values as low as 10^{-5} can be measured in this instrument (from Spruit).[406]

sliced into small, uniform pieces (about 1 mm long), placed into the cuvette, and slightly compressed by a plunger until the volume occupied by each individual sample is the same. Since metabolic changes occur in phytochrome *in vivo* at room temperature it is usual to keep the tissue at 1–2°C for the whole of the preparation and measurement procedures.

Other problems are associated with the measurement of relative phytochrome levels, one of the most troublesome being the existence of substances which exhibit unrelated spectral changes, such as protochlorophyll, or others which contribute massively and overridingly to the optical density in the red or far-red regions, such as chlorophyll. The presence of chlorophyll presents a currently insuperable problem to the measurement of phytochrome in fully light-grown plants and, except for variegated or colourless tissues, it is only possible to detect phytochrome after removal of all masking substances, i.e., after extraction from the tissues. Another source of trouble in the routine *in vivo* estimation of phytochrome is the existence of wavelength-dependent and wavelength-independent shifts in tissue absorbance. Such shifts occur rapidly, although often in a continuous steady manner, and are probably due to changes in the relative disposition of cellular components due to the absorption of energy from the actinic red and far-red sources. These changes may lead to differences in light-scattering and refractive indices, and thus bring about the observed shifts. Most investigators make measurements of $\Delta(\Delta A)$ from a series of red and far-red actinic irradiation treatments, and thus are able at least to correct for wavelength-independent shifts. A continuous drift can be discerned in the readings in Table 4.1.

4.2 Isolation and purification of phytochrome

The first isolation of photoreversible phytochrome in a cell-free preparation was achieved by the use of the classical methods of protein chemistry.[65] Since then, the

Table 4.2 The Siegelman and Firer[387] procedure for isolating phytochrome.

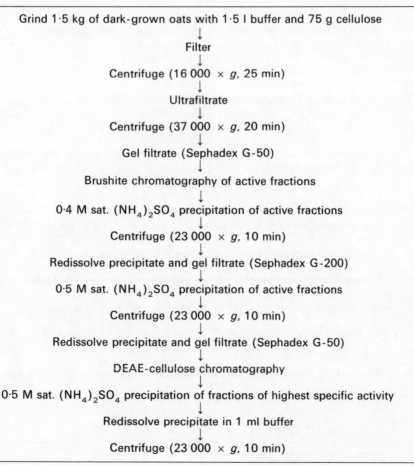

Grind 1·5 kg of dark-grown oats with 1·5 l buffer and 75 g cellulose
↓
Filter
↓
Centrifuge (16 000 × *g*, 25 min)
↓
Ultrafiltrate
↓
Centrifuge (37 000 × *g*, 20 min)
↓
Gel filtrate (Sephadex G-50)
↓
Brushite chromatography of active fractions
↓
0·4 M sat. $(NH_4)_2SO_4$ precipitation of active fractions
↓
Centrifuge (23 000 × *g*, 10 min)
↓
Redissolve precipitate and gel filtrate (Sephadex G-200)
↓
0·5 M sat. $(NH_4)_2SO_4$ precipitation of active fractions
↓
Centrifuge (23 000 × *g*, 10 min)
↓
Redissolve precipitate and gel filtrate (Sephadex G-50)
↓
DEAE-cellulose chromatography
↓
0·5 M sat. $(NH_4)_2SO_4$ precipitation of fractions of highest specific activity
↓
Redissolve precipitate in 1 ml buffer
↓
Centrifuge (23 000 × *g*, 10 min)

protein nature of phytochrome has been unequivocally proven by its successful purification in large amounts from several plants.

The most popular plants for phytochrome extraction have been oats[197, 253, 308, 354, 359, 387, 391, 450] and rye,[75, 354] although the pigment has also been purified from maize,[183] barley,[183] and peas.[422] Full details of the pea phytochrome have not yet been published, and although phytochrome has also been isolated from the alga *Maesotenium*,[422] the liverwort *Sphaerocarpos*,[422] and the moss *Mnium*,[144] it would seem vitally important in future work for comparisons to be made between cereal phytochrome and that from a wide range of other species.

The purification of phytochrome is fraught with difficulties – it is present in cells in very low concentrations; it is unstable to conditions which oxidize sulphydryl groups; the Pfr form is very unstable especially in the presence of metal ions; and in certain tissues there appear to exist proteolytic enzymes which attack phytochrome specifically. In order to cope with these problems, a rather complex and time-consuming extraction and purification procedure is required, and since extensive facilities for the cultivation and processing of very large amounts of material are

Table 4.3 The Briggs procedure for isolating phytochrome (compiled from Gardner, *et al*.)[141].

Chilled and harvested oats (1 kg).
Grind in 1 l buffer: 50 mM *tris* + 0·7 per cent 2-mercaptoethanol
Extract in a Waring blender
↓
Centrifuge at *ca*. 3000 rev/min 15 min
↓ discard pellet
Make the supernatant 10 mM in $CaCl_2$
↓
Centrifuge at *ca*. 3000 rev/min, 15 min
↓ discard pellet
Add the sample to Brushite column (1 l brushite per kg fresh tissue) equilibrated with 2 column volumes 10 mM KPB*, pH 7·0 followed by 2 volumes 10 mM KPB + 0·7 per cent 2-mercaptoethanol, pH 7·0
↓
After sample addition wash column with 1 volume 10 mM KPB + 0·7 per cent 2-mercaptoethanol pH 7·0 followed by 1 volume 15 mM KPB pH 7·0
↓ step elute with 62 mM KPB pH 7·8
Crude phytochrome precipitated with 33 per cent $(NH_4)_2SO_4$
↓ centrifuge
Dissolve precipitate in buffer and dialyse for 1 h against same buffer
↓
Apply to DEAE-cellulose column equilibrated with 10 mM KPB pH 7·4
↓
Elute with a convex gradient of 500 ml 0·3 M KCl in 10 mM KPB pH 7·4 into 250 ml 10 mM KPB pH 7·4
↓
Active fractions concentrated and precipitated with 40 per cent $(NH_4)_2SO_4$ pH 7·8
↓ discard supernatant
Redissolve preciptate and dialyse against 10 mM KPB pH 7·5
↓
PARTIALLY PURIFIED PHYTOCHROME

---

* KPB ≡ potassium phosphate buffer.

necessary, only a small number of laboratories have been able to tackle phytochrome purification successfully.

The first large-scale purification procedure was worked up by Siegelman and Firer, in 1964,[387] and most methods since then have been largely based on their scheme. This scheme, which is described in Table 4.2 uses repeated and alternated applications of different purification methods in order to adequately separate phytochrome from other proteins, and Siegelman and Firer were able to achieve a sixty-fold purification from oat coleoptiles. Later workers, using modifications of this basic method were able to achieve greater purification values, e.g., Correl et al.[75] purified phytochrome two hundred-fold from rye shoots, Mumford and Jenner,[308] seven hundred and fifty-fold from oats. Recently, Kroes[253] has devised a less time-consuming procedure for the isolation of very large amounts of phytochrome from oats, but this still takes two people $3\frac{1}{2}$ days to complete.

The purest phytochrome yet isolated was prepared by Briggs and colleagues in 1972[141, 354, 355] from oat and rye seedlings using the flow sheet in Table 4.3. These preparations have been shown to be single homogeneous proteins by acrylamide gel electrophoresis, and studies of their properties have led to a significant increase in our knowledge of phytochrome chemistry, as is described in detail below.

Phytochrome, therefore, is a protein. But, in order for a protein to absorb visible light, it must be associated with a highly conjugated, non-proteinaceous prosthetic group. Light-absorbing prosthetic groups are known as *chromophores*, and in this case the chromophore must be capable of existing in two forms, one of which predominantly absorbs red light, and the other far-red light. Our understanding of the properties of the protein moiety and the chromophore, and of the changes which take place in both upon photoconversion, form the themes of the next three sections.

4.3   The protein moiety

(a) Molecular weight

The molecular weight of purified phytochrome varies with the source of the preparation and the way in which it was purified. In Table 4.4, we can see that the reported molecular weights range from 16 000 to 260 000 daltons. It is unreasonable to suppose that this variability is a true representation of the nature of phytochrome within

Table 4.4 The published molecular weights of phytochrome.

Plant source	Method used	Molecular weight ($\times$ 10³ daltons)	Reference
Oats	Velocity sedimentation	90–150	Siegelman and Firer[387]
Oats	Molecular exclusion chromatography	55–62	Mumford and Jenner[86]
Oats	Molecular exclusion chromatography	55	Kroes[253]
Oats	Molecular exclusion chromatography	80 and 180	Briggs et al.[53]
Oats	Velocity sedimentation	16, 19, 120	Walker and Bailey[449a]
Rye	Velocity sedimentation	42, 180	Correl and Shropshire[75]
Pea	Velocity sedimentation	112, 260	Walker and Bailey[450]

these tissues, especially when phytochrome isolated from one plant, oats, can show molecular weights as different as 16 000 and 180 000 daltons. We are forced to conclude therefore, that the source of the variability lies in the methods used, and that either the phytochrome was differentially degraded in the different laboratories, or that there is some property of phytochrome which would cause it to have different apparent molecular weights when different ways of assessing molecular weight are employed.

Recently, a thoroughgoing attempt to rationalize this question has been made by Briggs and colleagues.[53, 328, 354, 355] Using the isolation procedure outlined in Table 4.2, they prepared pure phytochrome from oat and rye seedlings and proved that each was a single homogeneous protein, by ultracentrifugation, by exclusion chromatography on Sephadex gels, by Ouchterlony plate immunodiffusion, and by electrophoresis on acrylamide gels containing the detergent sodium dodecyl sulphate. In each case a single protein band was seen, and the molecular weights obtained were around 60 000.

If unpurified preparations were separated by Sephadex chromatography, on the other hand, two fractions could be seen, one with an apparent molecular weight of about 180 000 and the other of 60 000 daltons. Moreover, the heavy phytochrome disappeared rapidly on storage to give increased amounts of the lighter moiety. From this and other work, it was deduced that a proteolytic enzyme was being copurified with the phytochrome and that it had a specific action on phytochrome.[328] It was fortunately found possible to inhibit the protease with phenylmethylsulphonyl-fluoride (PMSF) and if the early stages of the preparation were done rapidly in the presence of PMSF, enriched amounts of the heavy phytochrome were obtained. In further work, rye only was used since there appeared to be less of the protease present.

The molecular weight of the highly purified rye phytochrome was examined by several methods, and it became clear that the value obtained depended on the procedure used. The estimation of protein molecular weight is not a simple matter, one of the major difficulties being that certain methods only give useful results if the protein molecule is truly globular, i.e., is isodiametric. Procedures such as velocity sedimentation in the ultracentrifuge, and molecular exclusion chromatography on Sephadex, or similar gels, are markedly shape-dependent. Equilibrium sedimentation, however, in which the protein is spun in a density gradient until its position is equilibrated with its density, is virtually shape-independent. Another shape-independent method is electrophoresis on acrylamide gels in the presence of sodium dodecyl sulphate. In this case, the protein is completely denatured by the detergent and migrates as a linear polypeptide chain, the separation being due to the molecular sieving of the acrylamide gel. These gels can be calibrated with proteins of known molecular weight and are highly accurate. A comparison of the results obtained with highly purified heavy rye phytochrome, using these different methods is given in Table 4.5.

Briggs et al.[355] conclude from this data that the extracted phytochrome probably has a molecular weight of ca. 120 000 daltons and is not a globular protein. Since the equilibrium sedimentation procedure gave an exactly double molecular weight, it is suggested that this represents a favoured dimerization of the individual non-denatured molecules, and one that could conceivably exist in vivo. The 120 000 molecular weight component, when mixed with a crude preparation is degraded by the protease to units of 60 000 molecular weight, and it is suggested that the protease acts highly

Table 4.5 The molecular weight of highly purified rye phytochrome as measured by different procedures (Briggs et al.[354, 355]).

Method		Molecular weight daltons
Velocity sedimentation	Shape-dependent	180 000
Molecular exclusion chromatography	Shape-dependent	375 000
Equilibrium sedimentation	Shape-independent	240 000
SDS-gel electrophoresis	Shape-independent	120 000

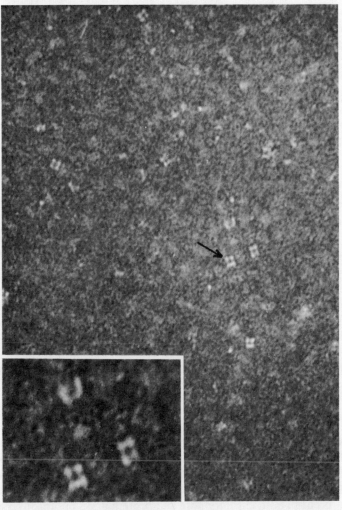

Fig. 4.5 Electron micrograph of highly-purified large rye phytochrome precipitated on a grid and negatively stained ($\times$ 250 000).Inset shows higher magnification ($\times$ 500 000) to reveal the 'double-dumbbell' appearance of the preparations. (Photograph generously provided by W. R. Briggs of a sample prepared by J. M. Mackenzie.)

specifically (at least initially) to cut the native molecule in half. The appearance of highly purified phytochrome when dried and viewed in the electron microscope (Fig. 4.5) suggests that the native 120 000 molecular weight subunit may be dumbbell in shape.

(b) Amino acid analysis

Several groups have prepared hydrolyses of purified phytochrome and analysed their amino acid contents. It would not be useful here to give full details of these analyses, but it is of interest that, even with very pure phytochrome, significant differences exist between different species. For example, Rice and Briggs[355] have shown that rye and oat phytochrome exhibit marked differences in their amino acid composition even though they appear to be confluent on immunodiffusion plates (see below). There are also very large differences, reported by different workers, for the amino acid compositions of phytochrome from the same species (compare Walker and Bailey,[450] with Mumford and Jenner[308], both using oat phytochrome). One generalization appears to be true, however; upwards of one-third of all the amino acids present are either acidic or basic (Table 4.6), indicating that phytochrome is a highly reactive, charged molecule.

Table 4.6 Summary of the amino acid analyses of highly purified oat and rye phytochrome (from Briggs et al.[354, 355]).

Plant	Total amino acids	Total basic	Total acidic	Cysteine
Rye (120×10^3 mol. wt.)	1140	133	232	26
Oat (60×10^3 mol. wt.)	588	86	113	12

(c) Immunological studies

The use of immunological techniques for the study of phytochrome has only very recently been tried, since it is futile to attempt to prepare an antiserum unless one has a single homogeneous protein. The use of this method, however, has enabled Rice and Briggs[354] to show that although oat and rye phytochrome must be almost identical in their antigenic sites, phytochrome from pea seedlings is significantly different. This recent work is most tantalizing as, if the protein moieties of phytochrome differ in different species, many interesting questions are raised concerning the molecular mechanism of phytochrome action. No doubt further comparative studies will soon be made on the apparent differences between phytochrome in different species.

4.4   The chromophore

The visible absorbance of phytochrome and its photoreversible changes are properties of a specific chromophore attached to a specific protein. For a full understanding of the mechanism of action of phytochrome it will be essential to determine the exact structure of the chromophore, of its linkage to the protein, and of the chemical changes that occur in both chromophore and protein upon photoconversion. In this section, we are concerned principally with the chemical structure as far as it is known, and with the methods used to elucidate it; the nature of the photoconversions is discussed later.

(a) Similarity of the phytochrome absorption spectrum to that of the bile pigments

As already stated, phytochrome has a highly specific absorption spectrum, and it was the observation that this absorption spectrum was very similar to that of certain algal pigments that led the Beltsville group to the first tentative identification of the nature of the chromophore. As early as 1950, nine years before the first isolation of phytochrome, Borthwick *et al.* had suggested that the pigment responsible for the photomorphogenic responses might be a linear tetrapyrrole, similar to the chromophore of phycocyanin.[40,322] Siegelman and colleagues,[391,392] later showed that the absorption spectrum of the purified phytochrome was, in fact, closely similar to those of C-phycocyanin and allophycocyanin (Fig. 4.6). All three substances have a sharp, long wavelength absorption maximum, varying from 610 nm to 660 nm, a pronounced shoulder around 570–600 nm, and a further absorption band in the blue-violet between 350–400 nm. The phycocyanins are known to be chromoproteins

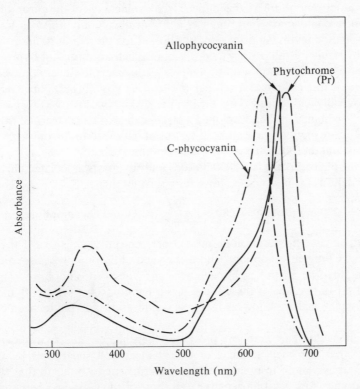

Fig. 4.6 Absorption spectra of the isolated chromophores of phytochrome, Pr (– – –); allophycocyanin (———); C-phycocyanin (–·–·–) (from Siegelmann *et al.*).[392]

with chromophores similar to the bile pigments of animals. When the phytochrome chromophore was separated from the protein (which, as will be seen below, is no easy matter) even greater similarities were observed between its absorption spectrum and those of the phycocyanin chromophores.

The tentative conclusion to be drawn from this work was that the phytochrome chromophore was likely to be similar to the bile pigments and all subsequent investigations have followed this line. As mentioned above, the bile pigments are linear tetrapyrrole molecules of the following basic structure:

The extent of conjugation along the chain, and the number of methyne bridges, varies from pigment to pigment. Although at the time of writing, the exact nature of all the substituents on the pyrrole rings of the phytochrome chromophore, and of the position of binding to the protein, remain incompletely resolved, it is nevertheless universally accepted that the chromophore is a linear tetrapyrrole.

(b) Separation of the chromophore from the protein

The greatest difficulty associated with experiments designed to determine the chemical nature of the chromophore has been the extreme reluctance of the chromophore and the protein to come apart. Siegelman et al.[392] achieved only a five to ten per cent yield by a method involving prior denaturation of the protein in trichloroacetic acid followed by refluxing the precipitated protein in methanol for 3–4 hours. Since the preparation of significant amounts of pure phytochrome is itself very time-consuming and laborious, it is clear that a ten per cent recovery of chromophore represents a very low overall yield indeed. This problem is strikingly illustrated by a calculation of Kroes,[253] who had earlier developed a large-scale extraction method for isolation of relatively pure phytochrome from 25 kg lots of oat seedlings. From such an extraction he could obtain 100–200 mg of protein containing phytochrome. If the molecular weight of the phytochrome is 60 000 daltons and that of its associated chromophore about 600 daltons (as it would be for a tetrapyrrole) then:

$$200 \text{ mg protein} \rightarrow 200 \times \frac{600}{60\,000} = 2\cdot0 \text{ mg chromophore}$$

If the phytochrome preparation is only 33 per cent pure, as is likely, this figure is reduced to 0·67 mg chromophore, and if the yield of chromophore isolation is only ten per cent, then we can at best hope to achieve 67 μg of chromophore from 25 kg of oat seedlings! In the face of these facts, it is a tribute to the sensitivity of chemical microtechniques that we know as much as we do about the phytochrome chromophore.

The small yield of chromophore prevented Siegelman et al.[392] from carrying out tests other than the measurement of absorbances and separation by thin layer chromatography. These tests were, however, sufficient to indicate that the phytochrome chromophore is a bilitriene (i.e., all three of the C-bridges between the pyrrole rings are methyne or —C≡ linkages) similar to, but not identical with, phycocyanobilin, the chromophore of allophycocyanin. This view is supported by the similar experiments also carried out on oat phytochrome by Kroes.[253] Siegelman et al.[386] later used these similarities to propose a full structure of the chromophore (see Fig. 4.13a).

(c) Degradation methods

Several other methods of chromophore isolation have been tried, based on techniques used successfully with other biliproteins, but none have to date yielded better results than the methanol method of the Siegelman group.[392] A different approach, however, has been very successful in providing information on the chemical nature of the chromophore, and this is degradation of the chromophore while still attached to the apoprotein. This method has been used successfully with other biliproteins by Rüdiger, who brought the methods into action against phytochrome in 1969.[363]

These methods, which are based on the chromic acid oxidation of pyrroles to yield maleimides with unchanged β-substituents (Fig. 4.7(a)) have several advantages for the study of phytochrome:

1. Bile pigments are quickly degraded at pH 1·5–1·7 at which porphyrins and chlorophylls (i.e., ringed tetrapyrroles) are relatively stable; it is therefore possible to characterize an unknown pyrrole pigment as a bile pigment purely by the kinetics of its degradation.

2. The yield of imides from the outer rings of a linear tetrapyrrole is good while, at the same pH, that from the inner rings is poor; instead dialdehydes are formed which become oxidized to imides at lower pH values (Fig. 4.7(b)); it is thus possible to distinguish between the inner and outer rings.

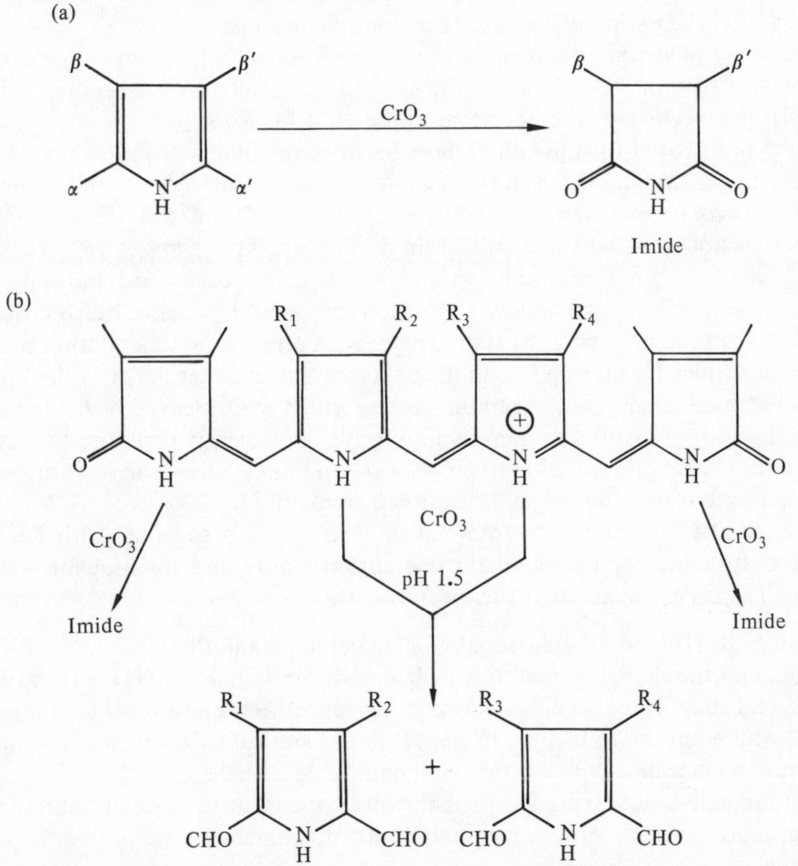

(a)

(b)

Fig. 4.7 (a) The chromic acid degradation of pyrrole substances. Note the β and β' substituents are not oxidized.

(b) The chromic acid oxidation of tetrapyrrole substances. Note the outer rings form imides while the inner rings form aldehydes.

3. If the degradation is performed at 20°C, ester linkages are stable, but these are saponified at 100°C; thus both hydrolytic and non-hydrolytic degradation can be carried out in the presence of the protein to provide evidence as to the binding of the chromophore to the apoprotein.

The first sample of phytochrome to be analysed by these methods had, unfortunately, lost its photoreversibility and thus could not be said to be truly representative of the native pigment,[363] but later work on freely photoreversible phytochrome has borne out most of the detailed results.[362]

The denatured form used first was greenish-yellow (as is Pfr) but could be converted to a blue form (corresponding to Pr) by cold acid, and reconverted to the greenish-yellow form by alkali. Both forms yielded identical products for three of the four pyrrole rings, upon chromic acid oxidation, as described below and shown in Fig. 4.8.

1. Ring II yielded the dialdehyde of haematinic acid at 20°C indicating that it was free from the protein, and an inner ring.

2. Ring III yielded haematinic acid imide only upon exhaustive oxidation at 100°C suggesting that it was covalently bound to the protein.

3. Ring IV from the blue and yellow denatured forms yielded an unidentified imide at 20°C suggesting it was also free from the protein.

With native phytochrome in either the Pr or Pfr form, only methylvinylmaleimide is formed. Native phytochrome also gave identical results for the products of rings II and III irrespective of whether it was in the Pr or Pfr form.

Only ring I proved to show differences between the blue and yellow forms of the denatured pigment, and between the denatured and the native states of the pigment. The blue form gave methylethylidenesuccinamide at 20°C and must therefore contain an exocyclic double bond and be free from the protein. The yellow form, on the other hand, gave methylethylmaleimide, with an endocyclic double bond, but only under hydrolytic conditions. The yellow form therefore must have ring I also covalently bound to the protein. Using the native pigment, however, no degradation products have been obtained from ring I, with the phytochrome in either the Pr or the Pfr form suggesting that there is a linkage to the protein which is not cleaved by the hydrolytic degradation. Brief alkali treatment will, however, sufficiently denature the pigment to allow the chromatic oxidation to proceed and in this case the products are identical to those formed from the original denatured preparation.

These results are shown diagrammatically in Fig. 4.8 together with Rüdiger's proposed structure for the phytochrome chromophore and its attachment to the protein. The salient features of this proposal are:

1. Rings II, III, and IV are probably identical in Pr and Pfr.

2. The chromophore is probably attached to the protein through ring III, probably via the acid substituent. (This conclusion has since been questioned by Briggs and Rice,[54] who point out that rings II and III have identical substituents, thus making it difficult to determine which is protein bound.)

3. In the native state, ring I is probably also bound to the protein, although the exact chemical identity of this ring and of its attachment to the protein is, as yet, unknown.

This proposal is not the only published model for the phytochrome chromophore (see also Siegelman *et al.*[386]), but it is supported by the strongest chemical evidence. Nevertheless, although it represents a remarkable achievement we are still in doubt

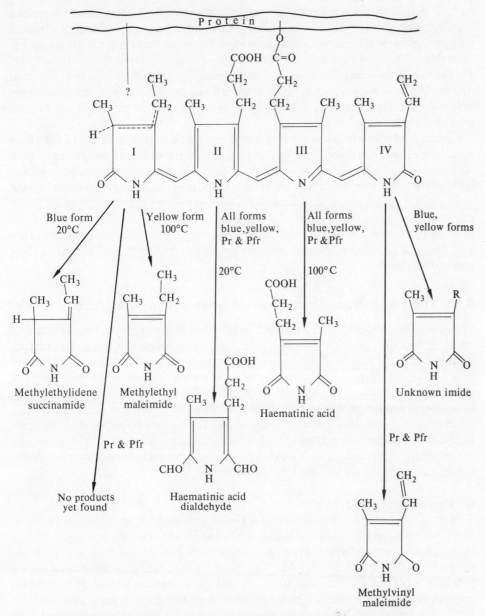

Fig. 4.8 The chromic acid degradation of phytochrome as performed by Rüdiger.[362]

concerning the true chemical differences between Pr and Pfr, knowledge of which is essential to our ultimate understanding of the way phytochrome acts.

(d) Chromophore number

The number of chromophores per phytochrome molecule is still not known for certain, although it has always been tacitly assumed that 60 000 molecular weight phytochrome contains only one chromophore. Correll et al.,[76] however, pointed out in 1968 that there is no evidence for this assumption. Their data, based on studies with rye phytochrome probably of 120 000 molecular weight revealed a much more complex picture. Kinetic data suggested the existence of two populations of Pfr reverting in darkness with different rate constants to two different populations of Pr. Furthermore, they detected four different absorbing regions at 580, 660, 670, and 730 nm; sodium dodecyl sulphate bleached all but the 580 nm band and destroyed photoreversibility, whereas glutaraldehyde removed the 730 nm band but had no effect on photoreversibility.

Correll et al.[76] constructed, from this information, a complicated model with four spectrally distinct chromophoric species (i.e., two forms of Pr and two forms of Pfr), which can interact by means of coupled oxidation–reduction reactions between the 580–660 nm and 670–730 nm pairs of chromophores. This hypothesis has not yet been confirmed by other studies and must remain conjectural until more direct evidence on chromophore number is obtained. At present, however, it would seem prudent to assume that each 120 000 molecular weight phytochrome molecule probably contains two chromophores, with the possibility that they may well prove to be spectrally different.

4.5   The phototransformation of phytochrome *in vitro*

With the availability of purified phytochrome, albeit at the expense of much time and effort, it becomes possible to investigate the physico-chemical changes occurring during the photochemical interconversion of Pr and Pfr. This has been a most fruitful area of study involving the application of a very wide range of techniques with the result that we can now be reasonably certain of the general principles underlying these changes as they occur in the test tube. Whether these changes occur in the same manner *in vivo* is another question and one which is not yet resolved; it would seem likely, however, that the processes within the plant are at least akin to those seen in the purified preparations. The *in vitro* changes are discussed here and the corresponding *in vivo* processes are left until the next chapter.

(a) Phototransformation kinetics

To date, all *in vitro* phototransformation kinetic experiments, in the absence of arte-facts, have shown the strict log-linear relationships expected of first-order photo-chemical reactions. A typical example, obtained by Everett and Briggs[114] with oat phytochrome, is shown in Fig. 4.9. This data, in fact, was published to correct a previous report by Purves and Briggs[344] that phototransformation of oat phyto-chrome gave curved kinetics which could be resolved into two linear components; later work had shown the result to be an artefact of the measuring instrument. The situation is not quite so clear, however, for *in vivo* studies, as is discussed on page 96; for the purposes of this section, it is sufficient to stress that phytochrome photo-conversions occur *in vitro* as typical first-order photochemical reactions.

(b) Absorption spectra, action spectra and photoequilibria

The first crude preparations of phytochrome showed that the pigment had the absorption bands predicted from the physiological experiments, i.e., preparations irradiated with red light had strong absorption in the far-red and *vice versa*. This is shown up in the visible colour difference between Pr and Pfr (Plate I). Detailed and

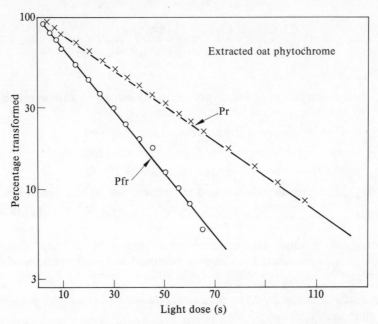

Fig. 4.9 Dose-response curves for the phototransformation of extracted oat phytochrome. Pfr = photo-transformation of Pfr to Pr; Pr = phototransformation of Pr to Pfr (from Everett and Briggs).[114]

accurate spectra were not available until later when more highly purified phytochrome became available.[62] The normal procedure is to pre-irradiate the phytochrome sample with an actinic source of far-red light so as to ensure all the pigment is in the Pr form, and then to scan the sample in the spectrophotometer. To obtain the absorption spectrum of Pfr, the sample is then irradiated with an actinic source of red light, and scanned again. Representative spectra obtained in this fashion are shown in Fig. 4.10(a).

This procedure, unfortunately, ignores a very important point, which as will be seen in chapter 5, is critical to our interpretation of the *in vivo* changes in phytochrome following irradiation. The absorption spectrum of Pr shown in Fig. 4.10(a) indicates that Pr absorbs very little light indeed at the wavelength of maximum absorption of Pfr, i.e., 730 nm. Pfr, on the other hand, exhibits considerable absorbance at 660 nm, the wavelength of maximum absorption by Pr, and also over a wide range of other wavelengths in the red region of the spectrum. Consequently, although 730 nm light will photoconvert virtually all Pfr to Pr, 660 nm light will only set up a photoequilibrium between Pr and Pfr, since both molecules will absorb and will be photo-transformed by light of this wavelength. This fact has several important implications, one of which is that the apparent absorption spectrum of Pfr in Fig. 4.10(a) is really the spectrum of the photoequilibrium mixture of Pr and Pfr set up by the red light used.

It is possible to calculate the actual absorption spectrum of Pfr if the proportions of Pr and Pfr in the photoequilibrium mixtures established by the actinic red light source are known. This calculation was first made by Butler *et al.*[62] in 1964 who showed that:

$$1 - [Pfr_\infty]_{665} = \frac{1}{1 + C_r/C_{fr}} \times \frac{A_{665\ min}}{A_{665\ max}}$$

and

$$\frac{C_r}{C_{fr}} = \frac{E_{725}A_{725\,max}}{E_{665}A_{665\,max}} \times \frac{(dA_{725}/dt)_{max\,665}}{(dA_{725}/dt)_{max\,725}}$$

where:

$[Pfr_\infty]_{665}$ = concentration of Pfr at photoequilibrium on irradiation with 665 nm light

$A_{665\,min}$ = absorbance of red-irradiated solution at 665 nm

$A_{665\,max}$ = absorbance of far-red irradiated solution at 665 nm

$A_{725\,max}$ = absorbance of red-irradiated solution at 725 nm

$(dA_{725}/dt)_{max\,665}$ = initial rate of absorbance change at 725 nm when far-red irradiated solution is irradiated with 665 nm light of intensity E_{665}

$(dA_{725}/dt)_{max\,725}$ = initial rate of absorbance change at 725 nm when far-red irradiated solution is irradiated with 725 nm light of intensity E_{725}

The absorbance values could be taken directly from the absorption spectra (Fig. 4.10(a)) and were as follows:

$$A_{665\,min} = 0\cdot59$$
$$A_{665\,max} = 1\cdot23$$
$$A_{725\,max} = 0\cdot64$$

The rates of change of absorbance at 725 nm due to irradiation at 625 nm and 725 nm were found to be proportional to the first-order rate constants K_{665} and K_{725}, which were measured to be in the ratio of 20:7. From these data, Butler et al.[62] were able to calculate that:

$$\frac{C_r}{C_{fr}} = \left(\frac{0\cdot64}{1\cdot23}\right)\left(\frac{20}{7}\right) = 1\cdot5$$

$$1 - [Pfr_\infty]_{665} = 0\cdot19$$

This means that with saturating red irradiation, at photoequilibrium, Pfr will represent 81 per cent of the total phytochrome and Pr will represent 19 per cent. Photoequilibria of phytochrome are commonly known as photostationary states and are expressed as [Pfr]/[Ptotal] either as fractions or as percentages. Thus, the photostationary state in saturating red light is 0·81 or 81 per cent. With this value, the true absorption spectrum of Pfr can be calculated. One such spectrum, calculated by Hartmann,[167] is given in Fig. 4.10(b).

Knowing the photostationary state value, it is also possible to determine the action spectrum for the photoconversions in both directions. In order to construct the action spectra, Butler et al.[62] irradiated phytochrome solutions that were either 100 per cent Pr or 81 per cent Pfr with various wavelengths of monochromatic light of known intensity. The conversion of Pr and Pfr was then measured as a function of time, followed by a determination of the photostationary state for each wavelength. From these data, estimates of ϕ, the quantum yield of conversion for each wavelength, can be derived. When ϕ is multiplied by the specific extinction coefficient, ε, for each

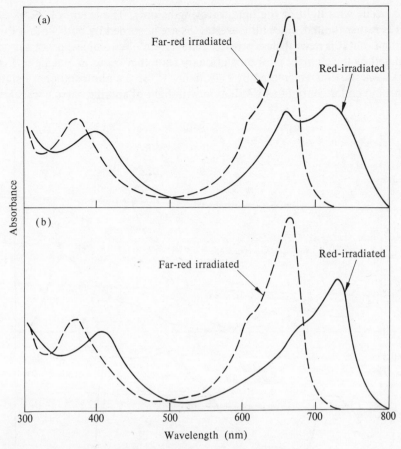

Fig. 4.10 Absorption spectra of purified phytochrome in the Pr and Pfr forms:

(a) direct measurements (reprinted with permission from Butler *et al.*[66], copyright American Chemical Society).

(b) corrected for incomplete phototransformation of Pr to Pfr (reprinted with permission from Hartmann).[167]

wavelength, the action spectrum of conversion is generated. This can be done for the conversions in both directions, and the resultant curves are shown in Fig. 4.11(a). It is seen that the wavelength of maximum efficiency for the conversion of Pr to Pfr is 665 nm, and that for the conversion of Pfr to Pr is 725 nm.

The action spectra and the calculations from which it is derived also provide another important fact. The ratio of quantum yields for the Pr to Pfr conversion and the Pfr to Pr conversion, i.e., ϕ_r/ϕ_{fr}, is 1·5.[62] Furthermore, the ratio of the extinction coefficient of Pr at its absorption maximum, 665 nm, to that of Pfr at its absorption maximum, 725 nm, or $\alpha_{r665}/\alpha_{fr725} = 1·55$.[62] These facts go a long way to explaining the repeated physiological observations that reversal of a red light-potentiated response requires up to three or four times as much far-red light energy (see page 31).

A further interesting feature of both the absorption spectra and the action spectra so far not discussed, is the differences between Pr and Pfr in the blue regions. These are sufficiently large for reversible photoconversions between different photostationary states to occur with light of the blue wavelengths only. The extent of the possible changes is rather limited, since all wavelengths are absorbed by both Pr and Pfr, to some extent, but it is nevertheless possible that certain physiological processes could be regulated by the changes in photostationary state that occur in blue light. Figure 4.11(b) shows the data calculated by Hartmann[167] for the photostationary state of phytochrome under continuous irradiation with light of specific narrow wavelength bands.

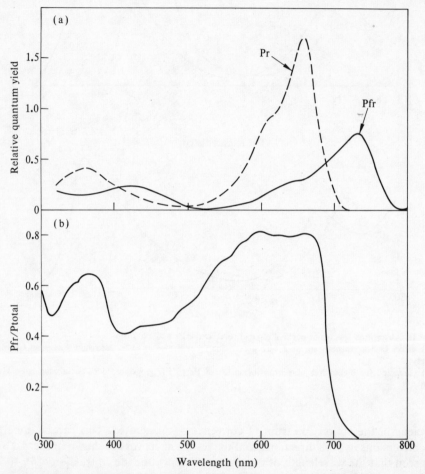

Fig. 4.11 (a) Action spectra of phytochrome photoconversions *in vitro* (reprinted with permission from Butler *et al.*).[62]

(b) Pfr as a proportion of Ptotal at photoequilibrium with various wavelengths (as calculated by Hart-Hartmann)[167] (reprinted with permission from Hartmann).[167]

(c) Evidence for intermediates in the conversions

When a solution of phytochrome in the Pr form is irradiated at 0°C with an intense, but very short, flash (5 microseconds) of red light, the absorption at 660 nm is lost immediately, but the absorption at 725 nm, characteristic of Pfr, does not fully appear until about 5 seconds after the flash. This type of experiment is known as flash spectroscopy and employs an oscilloscope to scan the changes in absorbance at a fixed wavelength as a function of time after the flash. The same experiment can be repeated at a series of different wavelengths, enabling difference spectra to be constructed for the changes that are occurring. Linschitz et al.[260] in 1966 were the first to carry out the above experiment and they observed that, although the 725 nm absorption peak did not appear directly upon the bleaching of the 660 nm peak, another peak at 695 nm was formed. This component, known as P_{695}, was formed in an immeasurably short time and it decayed after about 0·2 milliseconds to yield a further peak of absorption at 710 nm. P_{710} had a lifetime of about 0·5 milliseconds. Three more steps appeared to intervene between P_{710} and the final formation of Pfr. Linschitz et al.[260] concluded that P_{695} was formed directly from Pr by the red light, but was unlikely to be an excited state of Pr since its lifetime was too long. They were more inclined to think that P_{695} was an isomer of the chromophore, caused presumably by electronic redistribution. The subsequent steps were thought to be involved mainly in the re-orientation of the chromophore and the protein with respect to each other. Thus, the pathway of the Pr to Pfr photoconversion (at 0°C) was considered to be as follows:

$$\text{Pr} \xrightarrow{hv} P_{695} \longrightarrow P_{710} \begin{array}{c} \nearrow R_1 \searrow \\ \longrightarrow R_2 \longrightarrow \text{Pfr} \\ \searrow R_3 \nearrow \end{array}$$

Note: The authors make the rather unexpected proposal that the dark reactions involve several parallel, rather than sequential steps; this point will be discussed below.

The reverse transformation, i.e., Pfr to Pr, has been studied at 0°C in the same way. In this case, the flash of 725 nm light converts Pfr directly into a form absorbing maximally at 650 nm, and the intensity of absorption increases to its maximum value up to about 2 seconds, followed by further steps before the final formation of Pr:

$$\text{Pfr} \xrightarrow{hv} P_{650} \begin{array}{c} \nearrow R_1 \searrow \\ \quad\quad\quad \text{Pr} \\ \searrow R_2 \nearrow \end{array}$$

A refinement of the flash spectroscopy technique is to carry out the operations at low temperatures. If the temperatures are sufficiently low, the conversion intermediates are stabilized and more easily detected. With this method Linschitz's group[78] and Pratt and Butler[336a] have shown the formation of two other intermediates stable at low temperatures. The first formed intermediate has a peak at 692 nm (P_{692}) and this is converted in a fast dark reaction to P_{695}. P_{695} is stable up to $-70°C$ and can be reconverted back to Pr by actinic light. When the temperature is raised above $-70°C$, Pfr is formed slowly in several, temperature dependent steps. The last intermediate before Pfr is a bleached form with a very low absorbance centred around 650 nm (Pbl). These interconversions can be written as follows:

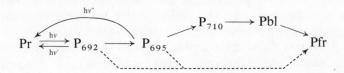

At 0°C Pbl is not seen at all, which could mean, either that the rate of its decay is greater than the rate of its formation, or that at the higher temperature a different route is used in which Pbl is bypassed altogether. This point becomes of some importance in the subsequent discussion of the chemistry of the photoconversions.

The existence of transformation intermediates with slow thermal decay constants can also be shown by modifications of ordinary spectrophotometric techniques. Briggs and Fork,[51] for example, irradiated phytochrome solutions with high intensity mixed red and far-red light while at the same time scanning their absorption spectra. The mixed red and far-red light causes rapid cycling of the pigment between the two stable forms Pr and Pfr and thus, once photoequilibria is reached, allows the accumulation of intermediates with slow thermal decay constants. Under these conditions, they observed the build-up of absorbance at 543 nm which decayed relatively slowly after the red/far-red light was removed. Kinetic analysis of the decay signals suggested that they represent the simultaneous independent and parallel decay of two species by first-order kinetics to Pfr. The same authors were able to show that exactly analogous intermediates build up in living tissues irradiated with mixed actinic red and far-red light.[52] This work is discussed with other similar findings in chapter 5 where *in vivo* phototransformation is considered.

The effect of low temperatures in slowing down the transitions also enables informative spectrophotometric experiments to be carried out. Kroes[253] has shown by spectrophotometry of phytochrome solutions cooled to $-196°C$ that Pr is initially photoconverted to P_{693}, confirming the results of Linschitz. When P_{693} is warmed to $-15°C$, Pfr is formed by a series of dark reactions. If the conversion is carried out in the reverse direction, the first formed substance has an absorption maximum at 663 nm which is rather broad; upon raising the temperature to $-15°C$, this band sharpens and intensifies to give the normal Pr peak at 665 nm.

All this information is probably most simply interpreted in the way proposed by Kroes.[253] The transformation of Pr to Pfr begins as the uptake of an amount of light energy which causes an isomerization of the chromophore to yield an unstable transition state (P_{693}). The extra energy of this state induces a local conformational change in the protein to produce a new complex between protein and chromophore, Pfr. The reverse transformation is energetically easier since the Pfr protein-chromophore complex is thought to be less stable; thus the lower energy quanta of far-red light is sufficient to drive the reverse conversion. This model is shown in Fig. 4.12.

(d) Evidence for a change in conformation of the protein

In 1964, Butler *et al.*,[66] at Beltsville, showed that Pr and Pfr differed in their sensitivity to denaturation treatments. If Pr was incubated with 5 M urea, a classical denaturing agent, the molecule was denatured to a form (designated Pr*) with a different absorption spectrum. Pr* was, however, stable and no further loss of the chromophore occurred. If, however, Pr* was converted to Pfr* by light, or if urea was added to phytochrome in the Pfr form, the chromophore deteriorated progressively with time. Further, Pfr was considerably more susceptible to degradation by proteolytic enzymes than was Pr. These results indicate that the protein in Pr is in a different, and more stable, conformation, than the protein in Pfr, implying, obviously, protein conformation changes associated with the phototransformations.

Briggs *et al.*,[57] however, were subsequently unable to find any differences between Pr and Pfr, in velocity sedimentation in sucrose gradients, electrophoretic mobility, behaviour on molecular sieve chromatography, and binding to and elution from calcium phosphate gels. They did, on the other hand, observe that Pfr was more labile during ammonium sulphate precipitation than Pr. Roux and Hillman[360] also obtained differences between Pr and Pfr which implied conformational changes in the protein moiety. Pr was more sensitive to glutaraldehyde fixation than Pfr,[359] and this was correlated with the finding that, out of a total of 27 lysine residues in highly purified 60 000 molecular weight phytochrome, 13 reacted with glutaraldehyde when Pr was fixed, and only 11 when Pfr was fixed.

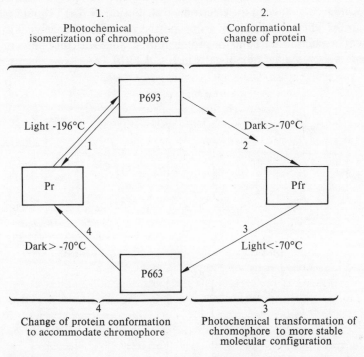

Fig. 4.12 Scheme for the sequence of events during the phototransformation of Pr to Pfr and *vice versa* as constructed by Kroes[253]

Hopkins and Butler[198] also observed minor differences between Pr and Pfr in ultraviolet difference spectra, complement fixation of Pr and Pfr antibodies, and circular dichroism. Finally, careful differential velocity sedimentation studies by Hopkins[197] showed that Pr has a larger sedimentation coefficient than Pfr by about $0 \cdot 1 s$ (Pr had an $s_{20, \omega}$ value of $5 \cdot 1$ and Pfr $5 \cdot 0$).

All these data are consistent with a protein conformational change accompanying the photoconversions. It seems likely, however, that the changes are not large, and may be restricted to the regions close to the chromophore.[54]

(e) The chemical nature of the photoconversions

As yet, the exact nature of the chemical changes that occur during the photoconversion of Pr and Pfr and *vice versa* are unknown, but several intelligent guesses have been made. The first complete proposal was published by Siegelman *et al.*[386] in 1968, based on a chromophore structure, that has since been shown to be unlikely to be wholly correct, by the degradation work of Rüdiger[362]. Siegelman's scheme is shown in Fig. 4.13(a), and although probably no longer tenable, has the advantage of providing an explanation for the marked changes in absorption since the proposed isomerization from Pr to Pfr extends the conjugated system of the molecule by three double bonds.

Rüdiger[362] has proposed an alternative scheme in which ring I is involved in a proton exchange with the protein (Fig. 4.13(b)). For this to happen, a proton acceptor group must be present on the surface of the protein in the vicinity of ring I and this must stabilize one of the mesomeric forms. A further necessity is that the photochemically induced change in the electronic distribution of ring I should only enable proton transfer when the acceptor group on the protein is in the vicinity, i.e., the arrangement in space of ring I and the protein must change with the photoconversions; this concept is shown diagrammatically in Fig. 4.13(b).

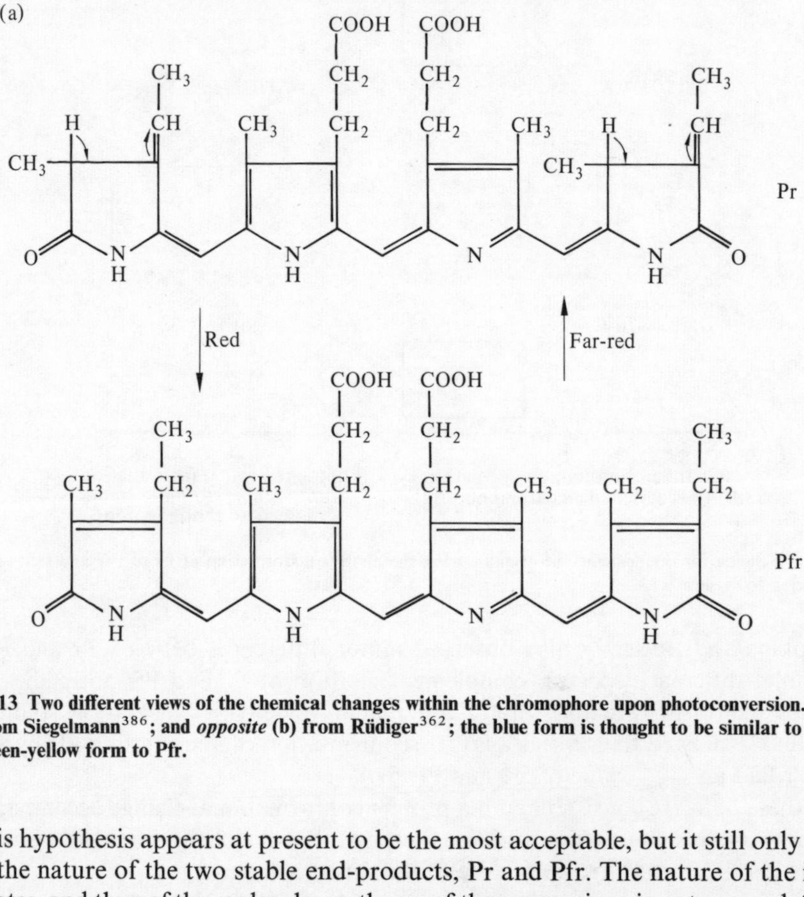

Fig. 4.13 Two different views of the chemical changes within the chromophore upon photoconversion. *Above* (a), from Siegelmann[386]; and *opposite* (b) from Rüdiger[362]; the blue form is thought to be similar to Pr and the green-yellow form to Pfr.

This hypothesis appears at present to be the most acceptable, but it still only deals with the nature of the two stable end-products, Pr and Pfr. The nature of the intermediates, and thus of the molecular pathway of the conversions is not covered. Some recent evidence, however, may give some indication as to what does, in fact, happen.

Burke et al.[59] investigated the circular dichroic spectra of phytochrome at low temperatures. With this technique it is possible to detect very slight changes in the asymmetry of complex organic molecules and it has been used with great effect to study the interaction of proteins and chromophores in several other chromoproteins.

Burke and his colleagues[59] found that the photoconversions in their samples of phytochrome could best be accounted for on the following scheme of intermediates:

$$\text{Pfr} \underset{+4°C}{\overset{-45°C}{\rightleftarrows}} \text{Pbl} \overset{h\nu\ -45°C}{\rightleftarrows} \text{Pr}$$

$$\text{Pfr} \underset{-70°C}{\overset{h\nu}{\searrow}} \text{A} \xrightarrow{-45°C}$$

All intermediates gave very large circular dichroic bands except Pfr suggesting that in Pfr the tetrapyrrole chromophore was probably more or less planar with very

little asymmetry. Pbl, however, a bleached intermediate probably identical to that found by other workers,[78, 260] gave an extremely large circular dichroic band, and the authors conclude that it is of a folded conformation. They suggest as one possibility that the Pr to Pfr conversion may begin with a *cis-trans* isomerization in the central bridge leading to the formation of a pseudoporphyrin; this would then immediately be followed in a dark reaction by a reverse *cis-trans* isomerization. The overall conversion would require the energy of one proton and thus could be achieved by the proton-donation scheme of Rüdiger.[362] The hypothetical isomerization is shown in Fig. 4.14.

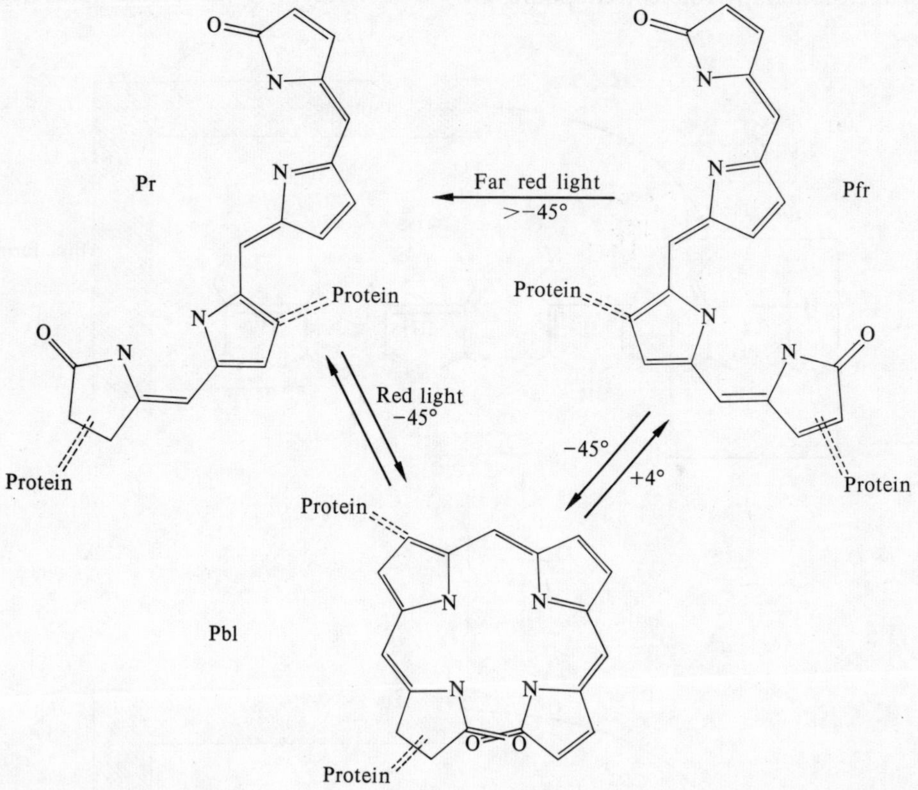

Fig. 4.14 Hypothetical scheme for the change in conformation of the phytochrome chromophore during phototransformation from Pr to Pfr at temperatures below 45°C. This scheme is based on the circular dichroism studies of Burke *et al.*[59] described in the text. The structures are drawn accurately to represent the actual distances between atoms. (Reprinted with permission from Burke *et al.*,[59] copyright American Chemical Society).

Thus we have a rather misty, but almost discernible, picture of the chemical nature of the phototransformations. To say they are complex is to be naïve, since all reactions of proteins are, almost by definition, complex. It is, however, intriguing to note the great similarities between the apparent photoconversions of phytochrome and the known photoconversion of the visual pigment rhodopsin. A glance back at Fig. 2.4 will illustrate the remarkable parallel of evolution that has occurred. As will be seen, the parallels between phytochrome and rhodopsin extend even further into the probable nature of the biological mechanisms of action.

4.6 Non-photochemical conversions of phytochrome *in vitro*

It was suggested above that the Pfr form of phytochrome is essentially less stable than the Pr form. This property allows the Pfr form to take part in non-photochemical reactions that do not have analogies with Pr. The basic observations are, that Pfr will slowly revert back to Pr in darkness, and will also, often more rapidly, change to a hitherto unidentified form which has lost photoreversibility. Both of these processes were originally discovered *in vivo* but later it was found that similar processes occurred *in vitro*. Much earlier interest in *in vitro* studies was centred on the phenomenon of reversion since reversion was, at one time, considered a likely contender as a timing mechanism in photoperiodism. More recent work has thrown doubt on the 'reversion as timer' theory and the *in vitro* studies are important now only in so far as they provide information on the chemical nature of phytochrome (see Prue[343] for a detailed treatment of this point).

Anderson *et al.*[4] showed that dark reversion of Pfr to Pr in purified oat phytochrome occurred as a single first-order reaction with a rate strongly dependent on

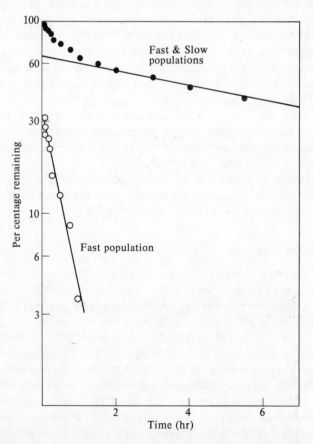

Fig. 4.15 The kinetics of dark reversion of purified high molecular weight rye phytochrome *in vitro* showing evidence for two populations. The upper curve gives the observed dark reversion of the preparation, the line representing the reversion rate of the slowly reverting population; the lower curve was calculated by subtracting the rate of the 'slow' population from the observed results, and represents the reversion rate of the 'fast' population (from Pike and Briggs).[328]

both temperature and pH. For example, the half-life of Pfr at pH 7 and 25°C was 9 hours, whereas at pH 6·2 it was 3·2 hours. In contrast to these results, Taylor[421] obtained complex reversion curves with oat phytochrome. Correl et al.[76] with rye phytochrome reported reversion curves which could be resolvable into two first-order components. Similarly Pike and Briggs[327] using highly purified 120 000 molecular weight rye phytochrome, obtained reversion curves resolvable into two first-order components (Fig. 4.15). The high molecular weight phytochrome showed an extremely rapid reversion rate even at 5°C, but if 60 000 molecular weight rye phytochrome was used, although two component curves resulted, the overall rate was very much slower. It seems that the many differences in the literature regarding rates of reversion could be explained by different degrees of protein degradation.

The observation of two component reversion rates suggests that two populations of phytochrome might exist, at least as far as reversion is concerned. This is an intriguing possibility since, as stated above, virtually all investigations of phototransformation kinetics to date have shown strict log-linearity. Thus, one population of phytochrome molecules as regards phototransformation can apparently follow two kinetically distinct reversion routes. Neither the explanation of this curious behaviour, nor its possible physiological importance, is at present known.

Other work on purified oat phytochrome[308] has shown that several reducing agents, including NADH, reduced ferredoxin and dithionite, dramatically increased the in vitro reversion rate. The increases were most marked with reduced ferredoxin which stimulated the rate of reversion about eight hundred-fold at a concentration of 10^{-6} M. These results are intriguing in relation to those of Klein and Edsall,[242] who showed that, in vivo, oxidizing agents mimic the effects of far-red light and reducing agents that of red light; the two effects, however, are in completely opposite directions. Also, phytochrome would need to be located in chloroplasts for ferredoxin to have any effect in vivo (see Section 5.1).

Finally, it should be noted that all the information on dark reversion of purified phytochrome has been obtained using phytochrome extracted from tissues in which in vivo dark reversion does not occur (see chapter 5).

4.7 Conclusions

The study of the chemistry and physics of isolated phytochrome over the last decade has been the most rewarding of all the investigations devoted to photomorphogenesis. We now know that phytochrome has a molecular weight of about 120 000 daltons, or possibly a multiple of this figure within the living cell, is probably not a globular protein, and is likely to have two chromophores per 120 000 molecular weight unit. The chromophores may not be identical with respect to their intramolecular environment, which could possibly account for the apparent multi-component dark transformation kinetics. The chromophores are bilitrienes whose chemical structure is almost fully known although much speculation still centres around the chemical nature of the photoconversions. The involvement of small protein conformational changes in the photoconversion is certain. Thus, photon absorption by the chromophore appears to cause a change in the electronic configuration of the chromophore (perhaps by cis-trans isomerization) which in turn is accompanied by a change in the tertiary conformation of the protein moiety. As pointed out already, this overall behaviour, although quite different chemically, has many striking similarities to the sequence of events involved in the photoconversions of the visual pigment rhodopsin.

This chapter has been an introduction to the properties of isolated phytochrome; much more is known (see Briggs and Rice[54] for a recent review) and much is yet to be discovered. What is already known, however, correlates remarkably well with the phenomenology of photomorphogenesis in higher plants as will be seen in the subsequent chapters, and this leads one to hope that further study of isolated phytochrome will provide information of great value to a fuller understanding of the mechanism of phytochrome action.

5

The biological action of phytochrome

The major outstanding question concerning phytochrome is the nature of the molecular mechanisms through which it exerts its influences on cellular development and metabolism. In this chapter, we explore the state of present knowledge on the behaviour and properties of phytochrome within the plant, culminating in a speculative discussion of the various hypotheses of the immediate mechanism of action of phytochrome. Integration of these hypotheses into a consideration of the control of photomorphogenesis is postponed until the final chapter.

5.1 Distribution and location of phytochrome within the plant

(a) Intercellular distribution in the plant

The question of where phytochrome is located within the plant is very important, since if phytochrome is closely involved in the processes of growth and development, it would be expected to be concentrated in the regions where such processes occur at maximum rate; namely, the meristematic and extending regions. Studies of phytochrome localization were made possible by the availability of the dual-wavelength difference photometer (the Ratiospect), but the requirements of the *in vivo* spectrophotometric methods have meant that such studies have been restricted to etiolated or other non-green plant materials. The *in vivo* spectrophotometric approach can be augmented by the extraction of phytochrome from green tissues, but it is difficult to determine the confidence limits of such methods when used quantitatively.

Phytochrome has been found either spectrophotometrically, or by extraction, in all parts of the more highly evolved plants including roots, stems, leaf blades, petioles, cotyledons, coleoptiles, hypocotyls, vegetative buds, floral receptacles, inflorescences, and developing fruits.[179, 180, 251] In etiolated seedlings, it is in fact true that the highest phytochrome content per unit protein is found in the meristematic and elongating regions.[55] This is also true of other meristematic tissue, e.g., the cambial regions of parsnip root tubers are sites of high phytochrome concentration.[251] High concentrations of phytochrome have also been found in cauliflower 'head' tissue,[163] and in the inner regions of brussel sprout buds.[190]

The apparent correlation with growth activity has been further extended in investigations of etiolated barley and maize seedlings, where it was found that early seedling growth in darkness was associated with an increase in total phytochrome per seedling.[63] Thus, it seems clear that phytochrome is concentrated in the tissues which are responsible for the developmental reactions to light treatment.

(b) Intra-cellular localization

1. *Cell fractionation studies.* Several attempts have been made to determine whether

or not phytochrome is organized into specific sub-cellular organelles. The most common approach to this type of problem is to homogenize the tissue in a buffer, and then to separate different fractions of the total cell homogenate by differential centrifugation. The earliest experiments along these lines appeared to indicate that phytochrome was partially associated with the mitochondria, and partially completely soluble.[147] However, later investigations of the properties of purified phytochrome showed that its solubility in aqueous solution is strongly dependent on pH.[392] At pH values above 7·0, phytochrome is largely water-soluble and in fractionation experiments tends to be recovered in the final supernatant. At lower pH values, it seems likely that phytochrome may be precipitated onto the surface of cell organelles, and thus be recovered in spurious association with them. The possibility still exists, however, that phytochrome is quite loosely attached to organelles *in vivo*, and that even the mildest extraction techniques result in its solubilization, especially in slightly alkaline buffers.

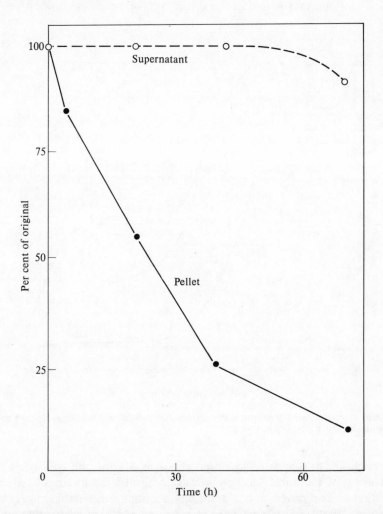

Fig. 5.1 Differential stability of particulate (pellet) and soluble (supernatant) phytochrome (after Rubinstein et al.).[361]

In spite of these difficulties, there is evidence that at least a fraction of the phytochrome present in etiolated plant cells may be associated with a membrane-rich fraction, and may also have different properties from the bulk of the phytochrome which is soluble in the cytoplasm. Rubinstein et al.[361] in 1969 working with oat seedlings found a small but consistent proportion of the total cell phytochrome (about five per cent) to be located in a fraction sedimenting between 1500 and 15 000 g. In this case, the binding of phytochrome to the membranous particles was not due to low pH since all operations were carried out at pH 7·4 and each fraction was washed in a buffer of pH 8·0. The only treatment which was successful in dissociating the phytochrome from the particulate material was washing with the non-ionic detergent Triton X-100. Furthermore, the pelletable phytochrome was very unstable at 0°C, upwards of 75 per cent being lost in 24 hours, whereas the supernatant was completely stable over that time period (Fig. 5.1). These results suggest that two fractions of phytochrome with different properties exist; one that is bound to membranes or organelles, and one that is free in the cytoplasm, or becomes so upon cell homogenization.

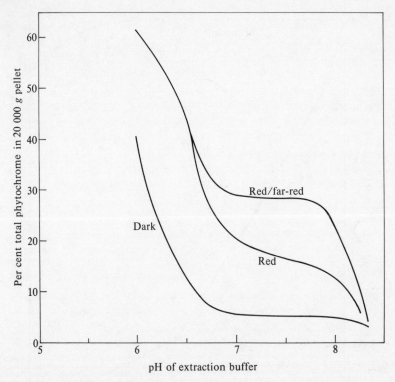

Fig. 5.2 **Effect of pre-irradiation treatment of maize coleoptiles and pH of extraction medium on pelletability of phytochrome at 20 000 g (after Quail et al.).[346]**

This conclusion is supported by observations by Marmé and colleagues[346] from the University of Freiburg. This group has confirmed the findings of Rubinstein et al.[361] that a small proportion of total phytochrome is pelletable at pHs between 6·8 and 8·0 at 20 000 g. Using maize coleoptiles and pumpkin hypocotyl hooks, they found that the percentage of total phytochrome found in the pellet varied with the irradiation conditions of the plant material immediately before extraction. Typical results are shown in Fig. 5.2 and Table 5.1, in which it is clear that red light increases

Table 5.1 Per cent pelletable phytochrome in extracts from maize coleoptiles and pumpkin hypocotyl hooks in the dark and following various irradiations with red (660 nm) and far-red (756 nm) light (from Quail et al.[346]).

Irradiation treatment	Per cent pelletable phytochrome	
	Maize coleoptiles	Pumpkin hooks
Dark control	5·6 ± 0·4	7·5 ± 0·4
5 min far-red	5·7 ± 0·5	7·4 ± 0·7
3 min red	15·9 ± 0·4	26·0 ± 1·3
3 min red + 5 min far-red	26·8 ± 0·6	16·9 ± 1·1
3 min red + 5 min far-red + 3 min red	15·6 ± 0·3	25·2 ± 1·7

the pelletability of phytochrome in both tissues. Curiously, however, subsequent far-red light increases even further the pelletability of the maize phytochrome, whereas the expected reduction in pelletability is seen with pumpkin hooks. These results are not easy to explain, but the original authors have attempted to do so on the basis of a speculative allosteric model involving the binding of phytochrome to specific sites on a membrane, whereupon photoconversion of Pr and Pfr (and back again in the case of maize) is thought to bring about changes in the binding sites, such that their affinity for phytochrome is increased. On this interpretation of the data, phytochrome is thought not to be a permanent membrane constituent, but to interact specifically with membranes upon photoconversion. Whether this view is correct remains to be seen; nevertheless, the data strongly suggest some sort of involvement of phytochrome with membranes.

2. *Microspectrophotometry*. Another approach to this problem has been the use of direct *in vivo* spectrophotometry of living tissue on microscope slides, using a microbeam of light. This sophisticated equipment employs a microscope through which the visible absorption spectrum of the part of the sample directly in the field of view can be scanned. Using this equipment, Galston[137] in 1968 was able to detect red/far-red photoreversible spectral changes when the microscope was focused on the nucleus of an oat coleoptile cell. The wavelengths of maximum spectral changes did not exactly coincide with the absorption maxima of phytochrome, but this may well have been due to absorption by other substances. The interpretation of these results is difficult, since as Briggs and Rice[54] have pointed out, the calibration of the instrument is arbitrary, and its sensitivity is unknown. Nevertheless, it seems incontrovertible that microbeams of red and far-red light when passed through nuclear regions cause reversible changes in light absorption. Such behaviour is normally accepted as a sufficient criterion for phytochrome involvement in physiological experiments and should be similarly accepted here. The actual spectral changes, however, seem likely to be due to substances other than phytochrome, since the observed changes are much greater than those calculated for the normal concentrations of phytochrome in cells.

3. *Evidence from polarized light beam experiments.* Evidence from indirect physiological experiments points to the localization of phytochrome in organized arrays in certain membranes, probably the plasmalemma. For example, when spores of the fern *Dryopteris filix-mas*[107] are irradiated with polarized red light in a certain energy range, the germ tube axis develops in a direction perpendicular to the electrical vector of the polarized light. This response, known as polarotropism, indicates that the photoreceptors must have a linear axis of absorption and, furthermore, must have their linear axes lined up and arrayed with respect to each other. The reason for this conclusion is shown in Fig. 5.3. Plane polarized light can only be absorbed by photoreceptor molecules which are oriented parallel, or nearly parallel, to the

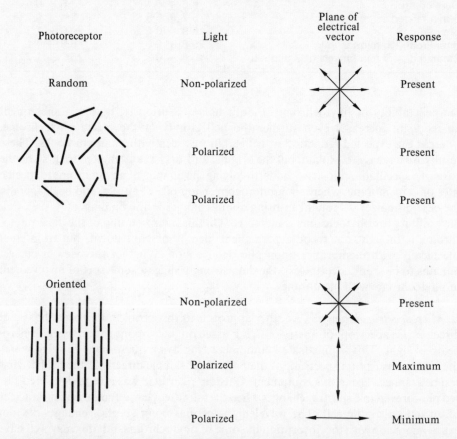

Fig. 5.3 Diagram to show how randomly arranged photoreceptor molecules will produce a response irrespective of the plane of polarization of the light (upper), whereas photoreceptors arranged parallel to each other can only absorb polarized light when its electrical vector is parallel to the direction of orientation of the molecules (lower).

plane of the electrical vector of the light. When the photoreceptor molecules are oriented randomly, or are free to move (as when they are in solution), on average the same number of molecules will be able to absorb light, no matter in what plane it is polarized; when the photoreceptor molecules are oriented parallel to each other, on the other hand, rotating the plane of polarization of the light will lead to a peak of absorption at the point at which the electrical vector is parallel to the linear axis of the oriented chromophores. Thus, whenever a biological photoresponse can be shown to be dependent on the plane of polarization of the light, it can be assumed that the photoreceptors are organized with respect to each other in such a way that their chromophores lie parallel. Such precise spatial organization of molecules can only occur if the molecules are either bound to each other directly, or connected indirectly through being attached to a flat sheet-like structure; in other words, a membrane. Polarotropic responses such as that described for *Dryopteris*, have also been shown in the liverwort *Sphaerocarpos donnellii*[411, 412] where similar conclusions as to the organization of the chromophores have been derived.

Further compelling evidence for the localization of phytochrome in (or near) the plasmalemma has come from the elegant work of Haupt and colleagues[172–175] on the light-mediated rotation of the chloroplast in the filamentous green alga *Mougeotia*. In general, chloroplast movement in higher plants is mediated by the absorption of blue light probably by a flavin pigment. *Mougeotia*, however, is an exception in that phytochrome also has a role in the response, as is also true for the unicellular green alga *Mesotaenium*. *Mougeotia* has a single flat, plate-like chloroplast which divides the vacuole into two, and which can rotate like a paddle within the cylindrical cell. Rotation is induced either by low-energy light, in which response phytochrome is the sole photoreceptor, or by high-energy light, a response which is much more complex and which probably involves two photoreceptor mechanisms.

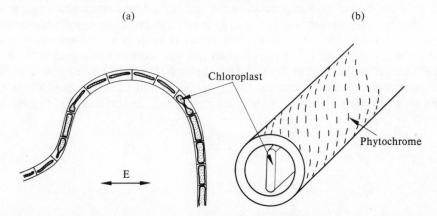

(a)   (b)

Fig. 5.4 (a) **Action dichroism in *Mougeotia*. The filament, originally with the chloroplasts in profile position, was irradiated with polarized red light normal to the plane of the paper; the double arrow E shows the plane of the electrical vector.**

(b) **Dichroic orientation of phytochrome (Pr) in a *Mougeotia* cell. The dashes show the direction of main absorption (after Haupt).**[172, 173]

For the low-energy response, Haupt first showed by the use of microbeams of red and far-red light that the chloroplast moved even if it was not itself irradiated. Thus, the particular molecules of phytochrome responsible for the initiation of rotation were shown to be outside the chloroplast, although, of course, this physiological evidence does not preclude the presence in the chloroplast of other molecules of phytochrome which are not involved in the rotation response. When a filament of *Mougeotia* was irradiated with plane polarized red light, it was found that whether

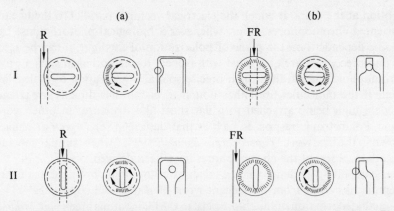

Fig. 5.5 Gradients of Pfr in a *Mougeotia* cell as established with a red or far-red microbeam. Phytochrome shown by surface parallel (Pr) and surface normal (Pfr) dashes. Left in each quarter is a cross-section with the initial state of phytochrome, the starting position of the chloroplast, and the localization of the microbeam; in the middle, expected phytochrome conversion and predicted chloroplast movement (arrows) are shown; right is a surface view with position of microbeam and response of chloroplast. Chloroplast starts in face (I) or profile position (II), the microbeam contains red (a) or far-red (b) irradiation (after Haupt).[173]

or not rotation occurred depended on the plane of polarization relative to the linear axis of the cell (Fig. 5.4a). From several such experiments, Haupt reasoned that the active phytochrome molecules must lie in a dichroic pattern with the main absorption vector of the chromophores parallel to the cell surface and oriented along a screw pattern (Fig. 5.4b). In addition, it was concluded from experiments with microbeams (Fig. 5.5) that the edge of the chloroplast moves away from a localized region of Pfr and towards a region of Pfr. Thus, rotation can only occur when a gradient of Pr to Pfr exists around the outer regions of the cytoplasm (Fig. 5.6). These results and conclusions are clearly best accounted for by assuming that the active phytochrome is fixed to some stable structure in the outer cytoplasm, and thus is able to exhibit a

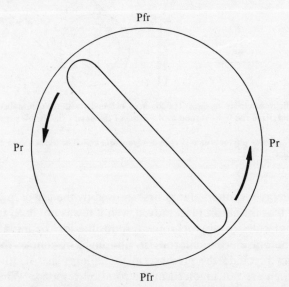

Fig. 5.6 The rotation of the chloroplast of *Mougeotia* away from Pfr and towards Pr (from Haupt).[172]

dichroic orientation and is prevented from diffusing throughout the cell. The best candidate for this structure is undoubtedly the plasmalemma, although there are other cytoplasmic structures which are sufficiently stable and oriented to allow the manifestation of a dichroic orientation of phytochrome.

Microtubules, which often lie parallel to the cell surface, and which are linearly oriented with respect to each other, could readily accommodate phytochrome molecules within their structure. Another possibility is the microfilaments present in the *Mougeotia* cell which seem to be directly involved in the movements of the chloroplast itself.[449a]

It thus seems clearly established that, at least in certain algae and ferns, a fraction of phytochrome is localized in specific molecular arrays in or near the plasmalemma. It is, of course, difficult to carry out similar experiments on multicellular plants, but one ingenious attempt indicates that phytochrome is similarly located in *Avena* coleoptile cells. In these experiments,[279] coleoptile cylinders were cut longitudinally and opened out into a flat sheet. A number of such 'opened' coleoptiles were then placed in the measuring beam of a Ratiospect and irradiated with actinic sources of polarized red and far-red light. When the red light was polarized at right angles to the longitudinal axis of the coleoptile it led to the photoconversion of 15 to 20 per cent more Pr to Pfr than when it was polarized parallel to the long axis. This suggests that phytochrome is also oriented into ordered arrays in coleoptile cells, which can again only be possible if it is part of, or is situated upon, a membrane.

4. *Immunocytochemical localization.* A further ingenious approach to the question of the intracellular localization of phytochrome is the use of immunocytochemistry.[337]

In this method, thin sections of plant tissues are reacted sequentially with a rabbit antiserum to phytochrome, a sheep antiserum to rabbit immunoglobulin, and a rabbit antiperoxidase-peroxidase complex. The result is that each phytochrome molecule binds rabbit antibodies, which, in turn, bind sheep antibodies, which finally bind the antiperoxidase-peroxidase complex. The peroxidase can then be visualized histochemically, indicating directly the location of phytochrome within the cells. Although phytochrome appeared to be concentrated in various cell membranes by this method, it was also observed in the cytoplasm. Interestingly, some cells appeared to contain no phytochrome at all.

In conclusion, all the available evidence at present indicates that at least part of the cellular phytochrome is located in some form of molecular array in, or near, the membranes of the cell; whether or not phytochrome is specifically located in the plasmalemma of multicellular plants is a question not yet fully resolved, although it is almost certainly so in those single-cell plants that have borne investigation. As will be seen below, the concept of phytochrome as a membrane agent has become of major importance to our thinking of how phytochrome might possibly act to regulate cellular activities.

5.2 Photochemical and non-photochemical conversions *in vivo*

(a) Photoconversion intermediates

As described in chapter 4, the photoconversions of Pr and Pfr, and *vice versa*, are complex reactions involving several different intermediates in each direction. There

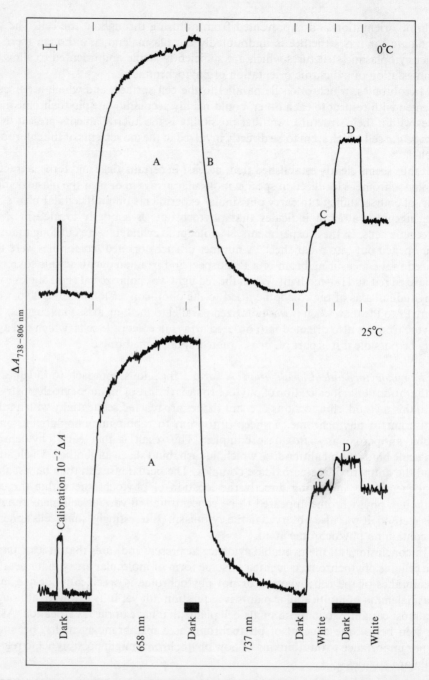

Fig. 5.7 Accumulation of phytochrome phototransformation intermediates *in vivo* under continuous white light. The traces represent recordings of differential changes in absorbance between 738 nm and 806 nm in *Amaranthus caudatus* samples. The conversion of Pr to Pfr under red light is seen in A and the reconversion to Pr under far-red light is shown in B. With white light only partial conversion to Pfr occurs (C); in subsequent darkness the further formation of Pfr is seen (D) representing the conversion of intermediates, in darkness, to Pfr. A greater proportion of Ptotal is held as intermediates under white light at 0°C than at 25°C (after Kendrick and Spruit).[231]

is as yet no compelling reason to believe that the reactions occur through very different pathways in living tissues, than in the test-tube. Briggs and Fork,[51,52] in fact, were able to show that similar long-lived intermediates accumulated during cycling of phytochrome in both tissue samples and pure preparations of the pigment, and thus it seems likely that the pathways of photoconversions are the same.

More recently, Kendrick and Spruit[231] have established that under high-intensity light, a significant proportion of the phytochrome present *in vivo* exists as intermediates. The method used was to determine the amount of Pfr in the tissue with the sample alternately exposed to actinic and measuring beams for periods of 1 ms, the presence of intermediates being detected by the increase in Pfr when the actinic beam was switched off (Fig. 5.7).

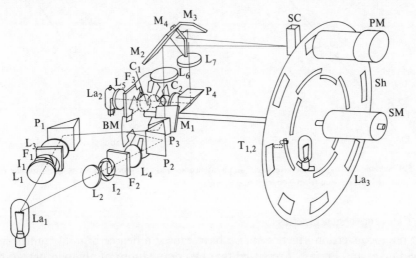

Fig. 5.8 A diagram of the quasi-continuous dual-wavelength spectrophotometer designed by Dr. C. J. P. Spruit and used to obtain the data of Fig. 5.7 (from Spruit).[407]

This work was made possible by the construction by Spruit[407] of a highly sensitive, dual-wavelength, quasi-continuous recording spectrophotometer. In this instrument, which is shown schematically in Fig. 5.8, the sample cuvette SC is irradiated by actinic flashes of 1 ms duration 400 times a second. The actinic source is La_2 and the beam chopping is by the 'butterfly mirror' BM rotated by the synchronous motor SM at 3000 rev/min. The differential transmission of the sample between two wavelengths (usually 738 nm and 806 nm for phytochrome) is measured during the intervals between successive actinic flashes using measuring beams derived from La_1, through the two filter assemblies F_1 and F_2. For phenomena with lifetimes, in darkness, of more than about 1·5 ms, the method forms a virtually continuous measurement of transmission during sample irradiation.

Kendrick and Spruit[231] have also shown that intermediate accumulation increases with irradiance, and is maximal in the 690–700 nm spectral region. Since naturally grown plants are exposed to high irradiances of white light (i.e., sunlight), these results are clearly of great importance when the function of phytochrome in the natural environment is considered (see chapter 7). On the basis of various low-temperature experiments, Kendrick and Spruit[232,234,407a] have proposed a new scheme for the interconversions between Pr and Pfr (Fig. 5.9). On this scheme, one of the intermediates is thought to be photochemically active in addition to Pr and Pfr (the only photochemical components in the other schemes), and absorption of light by this intermediate causes reconversion back to Pr. If this photochemical reaction takes place to any great extent, it must alter the photostationary state under continuous irradiation.

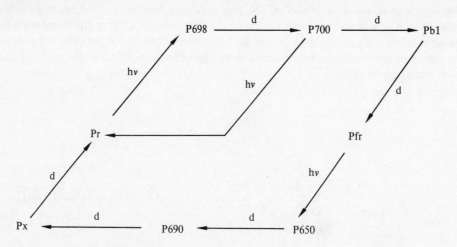

Fig. 5.9 Scheme for the photoconversions of phytochrome and its intermediates in *Amaranthus caudatus* (reprinted with permission from Kendrick and Spruit).[234]

(b) Photoconversion kinetics

The photoconversion kinetics *in vivo* have caused a degree of mild controversy. In 1968 Purves and Briggs[344] reported that two populations of phytochrome, differing in their rates of phototransformation, could be detected *in vivo* in oat, pea, maize, and cauliflower samples. These observations were carried out with the Ratiospect dual-wavelength difference photometer. Subsequently, the same group of workers reported that their earlier observations had been due to an artefact specific to the Ratiospect, and when experiments were carried out with other spectrophotometers, in all cases, simple first-order kinetics were observed.[115] Other workers had also found the phototransformations in both directions to exhibit simple first-order kinetics, and thus the possible existence of two populations of phytochrome was deemed unlikely.

In 1971, however, Marmé's group also reported deviations from first-order kinetics for phototransformation in pumpkin seedlings.[29] The data could be resolved into two first-order curves indicating that a 'fast' and a 'slow' population of phytochrome molecules existed together. Similar complex phototransformation kinetics were observed in three different spectrophotometers, and the proportions in the two populations could be changed by different pretreatments, suggesting that the results were not due to instrument artefacts. Later, however, Marmé and his colleagues[140] showed even these deviations were due to an artefact, although not an instrument artefact as such. When light passes through a cuvette containing plant material, there is necessarily a gradient of light intensity across the sample. Thus, phytochrome molecules in tissue near the light source will photoconvert more rapidly than similar molecules at the other side of the cuvette. When very thin slices of tissue were used, first-order phototransformation kinetics were obtained. This minor controversy serves to underline the importance of not making too rapid conclusions on the basis of kinetic evidence alone.

(c) Dark reversion

There are several ways in which the amount of total phytochrome, or the proportions of Pr and Pfr can be altered without the involvement of light. The first of these to be recognized is known as 'dark reversion' in which Pfr spontaneously, but slowly, reverts to Pr in darkness. The concept of the dark reversion of Pfr to Pr, developed principally from physiological investigations, which were apparently supported by early *in vivo* spectrophotometric observations of phytochrome in maize, in which plants treated with red light were placed in darkness, whereupon the Pfr slowly disappeared with the concomitant appearance of Pr.[64] During the dark transformations, however, a large part of the total spectrophotometrically detectable phytochrome was lost by the process of phytochrome destruction (see below). These experiments were complicated by two factors, the simultaneous loss of spectral photoreversibility by the phytochrome destruction process; and the, as then, unknown fact that even the purest sources of red light established only about 80 per cent of the total phytochrome as Pfr (see page 76). When these factors were fully taken into account, the apparent partial reversion of Pfr to Pr in maize was seen to be simply a measure of the 20 per cent total phytochrome remaining in the Pr form after the

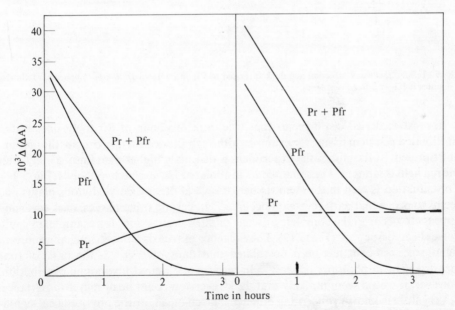

Fig. 5.10 The dark reversion artefact in maize coleoptiles.
(a) Actual data from the Ratiospect showing that after a red light treatment, Pfr disappears from the tissue within about 3 hours, while Pr plus Pfr at that time is approximately 30 per cent of original, assuming *all* Pr was initially converted to Pfr (after Butler *et al.*).[64]
(b) Diagram constructed to show actual events; since only 80 per cent of Pr can be converted to Pfr with red light, the total Pr plus Pfr at zero time cannot be measured in the Ratiospect; as the Pfr disappears (by destruction), increasing amounts of the remaining 20 per cent of Pr can be detected, giving rise to the impression that Pfr to Pr reversion is occurring in darkness.

initial red irradiation[63] (Fig. 5.10). Thus, the only dark transformation occurring in maize is the destruction of Pfr to a non-photoreversible form. This has also proved to be the situation in all other monocotyledonous plants investigated (e.g., oats, barley),[199] but only in certain dicotyledonous plants (e.g., *Amaranthus*).[230]

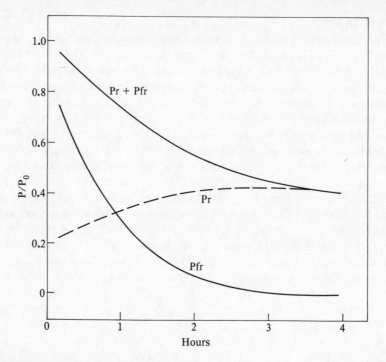

Fig. 5.11 **Simultaneous destruction and dark reversion of Pfr in etiolated** *Helianthus* **hypocotyls following 5 minutes red light (after Frankland).**[123]

In most dicotyledons, it seems that true dark reversion of Pfr to Pr does occur. In etiolated pea stem tissue, for example, although Pfr destruction occurs, the amount of Pr present at the end of the experiment is distinctly higher than immediately after the red light treatment. The same occurs in etiolated *Helianthus* hypocotyls (Fig. 5.11). This situation is seen in the stem tissues of several other dicotyledonous plants and seems to prove that dark reversion occurs.[199] In certain other tissues, dark reversion appears to occur in the complete absence of dark destruction, for example, in cauliflower head tissue[63, 64] (Fig. 5.12). The existence of true dark reversion has important significance, since it has been postulated that dark reversion is the basis of time measurement in photoperiodism.[178] In view of its limited taxonomic distribution, however, it would seem unlikely that dark reversion could be of major importance.

Very little is known concerning either the mechanism, or the physiological significance, of dark reversion. Although, as we shall see, the destruction process is sensitive to metabolic inhibitors, dark reversion does not appear to be. This has been used to show that dark reversion does in fact occur in dicotyledonous tissues. Furuya, *et al.*[134] treated pea stem segments with 2×10^{-3} M sodium azide, more or less completely stopping phytochrome destruction; under these conditions, apparent reversion was greater than in the non-inhibited controls. These results were interpreted to mean that reversion and destruction compete for the same Pfr molecules, and when destruction is prevented, reversion is increased. In maize, the same level of azide prevents destruction, but does not lead to any apparent increase in reversion, supporting the idea that dark reversion is limited in its distribution throughout the plant kingdom.

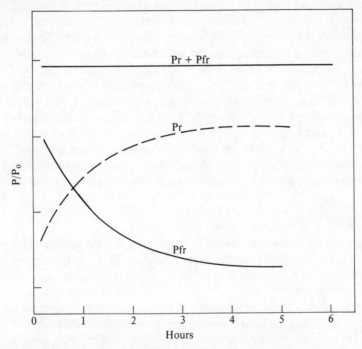

Fig. 5.12 Dark reversion of Pfr to Pr in the complete absence of Pfr destruction in cauliflower curd tissue after 5 minutes red light (after Butler *et al.*).[64]

A process formally similar to *in vivo* dark reversion also occurs in purified phytochrome preparations (see page 83). Whether or not this represents the same chemical change is uncertain, but it is interesting to observe that highly purified rye phytochrome reverted in darkness very rapidly, while dark reversion could not be detected at all in living tissues.[328] Another dissimilarity is that although *in vivo* dark reversion is markedly inhibited at low temperatures (i.e., *ca.* 5°C) reversion of the pure rye phytochrome proceeded unabated at that temperature. It is therefore still not possible to assess the physiological importance, if any, of dark reversion.

(d) Inverse dark reversion

The process of dark reversion described above is specific for Pfr, and it was thought for many years that the dark conversion of Pr to Pfr did not occur *in vivo* (neither was it known to occur *in vitro*). In recent years, however, observations on the germination behaviour of certain light-inhibited seeds forced several workers to the conclusion that such a conversion was possible, at least in newly imbibed seeds. This subject, as far as it concerns seed germination, is dealt with in considerable detail in chapter 6. It is sufficiently important as a general principle of phytochrome physiology, however, for it also to be outlined here.

Using a very sensitive spectrometer, Spruit and his colleagues were able to detect the presence of Pfr in completely dark-imbibed seeds of lettuce,[28, 30] cucumber,[408] and *Amaranthus caudatus*.[235] If this Pfr was removed by a single saturating exposure to far-red light, reappearance of Pfr occurred gradually over a period of about 20 minutes. During this time the total amount of phytochrome in the seeds did not change, and thus it was concluded that conversion of Pr to Pfr in the dark was taking place, a process which was termed 'inverse dark reversion' since it represents the opposite reaction to the long accepted dark reversion of Pfr to Pr. It must be stressed here that inverse dark reversion has not been found in any other tissues, nor under any other physiological conditions, and in any case appears to be thermodynamically unfavourable.[54] As will be seen, this correlates with other properties of the phytochrome detectable in newly imbibed seeds, which suggest that this fraction of phytochrome is physiologically distinct from the rest of the cell phytochrome. As discussed in chapter 6, inverse dark reversion appears to function, within those seeds in which it is found, as a mechanism for maintaining a certain level of Pfr necessary for germination, even in darkness, though why a photoreceptor should evolve a function specifically adapted to conditions in which light is absent is difficult to explain.

(e) Phytochrome destruction or decay

It has been shown already that the Pfr form of isolated phytochrome is considerably more reactive, and unstable, than the Pr form (chapter 4). This property has its parallel *in vivo* in the process commonly known as 'phytochrome destruction'. (The process is often known as 'decay' but since the term decay is apt to be ambiguous, the alternative of 'destruction' is favoured.[54]) Although Pfr breakdown *in vitro* has the same end result as *in vivo* destruction, there is no evidence that the same chemical processes are occurring. Phytochrome destruction, indeed, only means *the loss of photoreversible spectral changes*, and thus could be due to any type of chemical change, ranging from complete degradation of the molecule, to a small but irreversible change in electron distribution within the protein-chromophore complex, which results in the loss of photoreversibility. As yet, the actual mechanism of destruction is completely unknown. It is, however, known that destruction is dependent on metabolic activity, is virtually prevented at 0°C,[133] and is severely inhibited by anaerobiosis or metabolic poisons such as azide and cyanide.[63]

Most reports have stated that destruction follows first-order kinetics with a half-life of around $1\frac{1}{2}$ to 3 hours,[63] although one of the early investigations yielded zero-order kinetics.[336] These differences indicate, that in first-order kinetics destruction is proportional to the concentration of Pfr, whereas in the case of zero-order kinetics, destruction would be independent of Pfr concentration and be limited by some other factor. Whether this is an important distinction remains to be seen, but there is at least one species, *Amaranthus*, in which the relationship between destruction rate and Pfr concentration has been well established.

Although destruction is not light-dependent, the steady-state ratio of Pfr:Pr can be determined by light; e.g., red light maintains a photostationary state of 80 per cent Pfr while far-red maintains less than three per cent Pfr. By manipulating the wavelength of light in continuous irradiation experiments, Kendrick and Frankland[230] were able to show directly that the first-order destruction rate in *Amaranthus* seedlings was directly proportional to the Pfr:Pr ratio (Fig. 5.13). This is not always true for

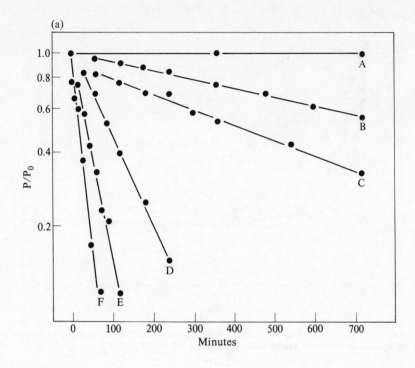

(a)

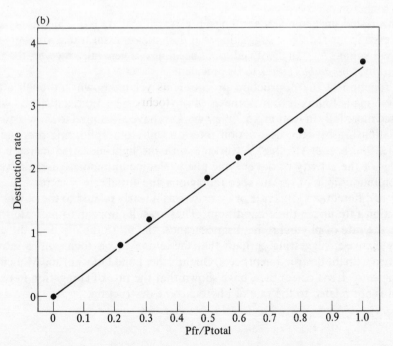

(b)

Fig. 5.13 The properties of the Pfr destruction process in *Amaranthus* seedlings;
   (a) The time-course of destruction (log plot) in darkness, A; and under different wavelengths of light,
B, 734 nm; C, broad-band far-red light; D, broad-band blue light; E, brief red light; F, continuous red light.
   (b) The linear relationship between photostationary state and the destruction rate; (the value of 1·0 for
Pfr/Ptotal is calculated from the loss of Pfr alone after brief red light). (Continued on page 102)

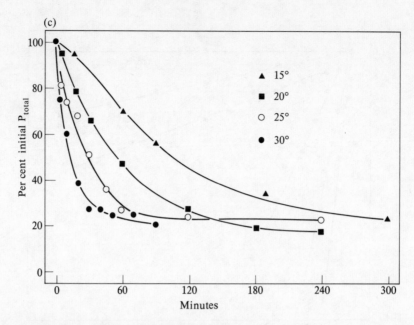

(c)

Per cent initial P$_{total}$

▲ 15°
■ 20°
○ 25°
● 30°

Minutes

Fig. 5.13 (continued from page 101) (c) The effect of temperature on the rate of destruction in darkness after a 5 minuted red light treatment (from Kendrick and Frankland).[230]

all plants tested since in maize seedlings, for example, there is a rapid disappearance of Pfr, even in far-red light, suggesting that destruction is saturated at relatively low Pfr/Ptotal values.[64, 72] In dicotyledonous seedlings in general, however, the pattern observed in *Amaranthus* seems to be prevalent.

The function of the destruction process is as yet unknown, although attempts have been made to link it to hypotheses of phytochrome action, a question which is discussed more fully in section 5.5. Other workers have attempted to show a quantitative relationship between destruction rates and photomorphogenic processes. For example, Smith and Attridge[403] working with the light-mediated increase in pea seedlings of the activity of the enzyme phenylalanine ammonia-lyase, showed that with continuous light of various spectral regions, the initial rate of increase in enzyme activity is different, and this rate of increase was linearly related to the phytochrome destruction rate under these conditions. These results appear to indicate that the greater the rate of phytochrome disappearance, the more rapidly does the enzyme activity increase, suggesting perhaps that the action of phytochrome is intimately associated with the destruction process. On the other hand Dooskin and Mancinelli,[88] working with *Avena* coleoptiles, have shown that the rate of elongation in response to light is not related to the rate of phytochrome destruction.

(f) Apparent resynthesis of Pr

When etiolated pea seedlings are irradiated with a short period of red light, and then returned to darkness, the total phytochrome level is rapidly reduced by the process of destruction and gradually stabilizes at about 40 per cent of the initial level. However, if repeated red light treatments are given at 90 minute intervals, the total phytochrome is reduced over a period of 8 hours to less than ten per cent of the initial value.[72] Subsequently, in darkness, total phytochrome level is seen to rise slowly for several hours. Similar results are obtained in seedlings kept in continuous red light (maintaining Pfr/Ptotal at 80 per cent and maximum destruction) for 8 hours, followed by darkness, as seen in Fig. 5.14. The deviation from first-order kinetics is here taken to indicate that when phytochrome content falls below a certain level in

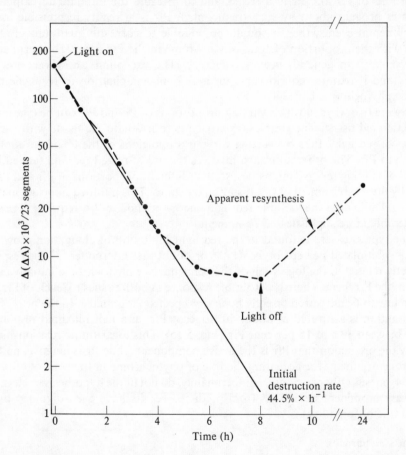

Fig. 5.14 Apparent resynthesis of Pr in pea epicotyls after loss of 90 per cent of Ptotal by destruction (from Clarkson and Hillman).[72]

peas a counterbalancing process of phytochrome synthesis begins. Experiments with inhibitors of protein synthesis appear to indicate that protein synthesis is involved, and thus it seems reasonable to conclude that the increase in photoreversibility is due to actual synthesis of new phytochrome molecules. The synthesis, of course, does not have to be occurring in those same cells which have lost their original phytochrome, and indeed it seems likely that the increases are due to the steady formation of phytochrome in new cells. Recently, Quail *et al.*[345] have demonstrated the *de novo* synthesis of phytochrome both in germinating seeds and under conditions of apparent resynthesis by the deuterium oxide (D_2O) density-labelling method.

5.3 Correlation of phytochrome with developmental responses to light treatment

Considerable effort has been concentrated on attempts to correlate the developmental responses to light treatment with the behaviour of phytochrome in the same tissues. The procedure employed in these experiments has been to establish known initial proportions of Pfr and Pr in a tissue, and to measure the ultimate developmental responses brought about by such treatment. If Pfr is in reality responsible for the developmental events, then it should be possible to correlate quantitatively these events with the proportion of total phytochrome in the Pfr form. The correlations achieved have, in general, been satisfactory. The exceptions, however, are very striking and it requires considerable ingenuity and imagination to reconcile them with the phytochrome concept.

There are two ways in which varying proportions of Pr and Pfr can be established in an etiolated tissue. The simplest approach is to irradiate the tissue with varying energies of red light, thus converting varying proportions of the Pr of the etiolated seedling to Pfr. The other approach involves the use of mixed red and far-red light sources of varying relative intensities, so that different proportions of Pfr and Pr are set up by the establishment of photostationary states. This photoequilibrium method is more reliable than the variable red light energy method, but it requires the use of considerably more sophisticated equipment.

In the typical cases, as found in the red light inhibition of elongation growth of sections of etiolated pea epicotyls,[191, 192] or of oat first internodes,[268] the response is directly related to the logarithm of the percentage of phytochrome initially established in the Pfr form. There is detectable response in both tissues to levels of Pfr that are too low to be measured directly by *in vivo* spectrophotometry. In the pea,[191, 192] the inhibition is saturated at about 50 per cent Pfr, and half maximal response is found between 10 and 15 per cent Pfr (Fig. 5.15). This logarithmic relationship fits exactly the prediction that Pfr is the active component while Pr is inactive, and this is a strong argument for accepting the role of phytochrome in the control of photomorphogenesis. Other responses, unfortunately, do not fit the hypothesis so elegantly. Two very prominent deviations from predicted results have been dignified by the special title of 'paradox'.

(a) The *Zea* paradox

The sensitivity to blue light of the 'first-positive' phototropic response in maize coleoptiles (see chapter 3) is markedly decreased by a pre-irradiation with red light. This effect is saturated at very low energies of red light, energies which do not lead to the formation of sufficient Pfr for detection by *in vivo* spectrophotometry. In fact, the energies required to produce sufficient Pfr for detection (i.e., *ca.* three to five per cent of Ptotal) are at least ten times greater than the energies required for saturation of the depression of phototropic sensitivity[71] (Fig. 5.16). Moreover, Briggs and Chon[50] have shown that the red light effect can be reversed by broad-band far-red

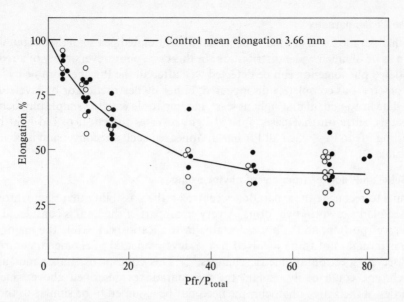

Fig. 5.15 Logarithmic relationship between the photostationary state of phytochrome (maintained for 15 minutes) on the subsequent extension growth of pea stem segments (after Hillman,[191] corrected for incomplete conversion of Pr to Pfr).

light, which is known to establish a photostationary state of between 3·5 and 5 per cent Pfr. Thus the response can be initiated by red light energies sufficient only to form less than three cent Pfr, while it is negated by far-red light which establishes more than 3·5 per cent Pfr. These data are obviously inconsistent with the simple hypothesis of phytochrome mediation.

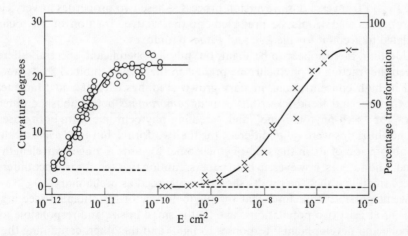

Fig. 5.16 The 'Zea paradox'. The curves are dose-response curves for phytochrome transformation (right) and the alteration of phototropic sensitivity (left) in maize coleoptiles. The dotted lines are dark controls. Note that a hundred times as much light energy is required to cause detectable Pr to Pfr conversion in the Ratiospect than is necessary to saturate completely the decrease in phototropic sensitivity (from Briggs and Chon).[50]

(b) The *Pisum* paradox

The other paradox was originally observed with etiolated pea stem tissues but is now known to be of much wider distribution. In this case, photoreversibility of a red light potentiated phenomenon can be detected well after all the Pfr, measurable by *in vivo* spectrometry, has completely disappeared, either by destruction or by reversion.[192] These data are again, at first sight at least, incompatible with the simple phytochrome hypothesis, since on that basis, far-red light reverses the effect of red light by reconverting Pfr to Pr – when all Pfr has disappeared such a reconversion is obviously impossible.

(c) 'Bulk' and 'active' fractions of phytochrome

Attempts to reconcile these paradoxes centre on the possibility that phytochrome in etiolated plants exists in two forms. A very small part of the total is considered to be the 'active' portion, and is not detectable in the Ratiospect, while the major part, which is thought not to be involved in the developmental phenomena, contributes the observed spectrophotometric changes.[193] This concept of 'bulk' and 'active' phytochrome could easily explain the two paradoxes described above, since the properties of the active phytochrome need not be assumed to be similar to those of the bulk fraction, except for the wavelengths of maximum activation.

What evidence is there that two populations of phytochrome exist each with different properties? It must be admitted that very little direct evidence on this question is available, but what does exist appears to be in favour of the concept. For example, it has been mentioned on page 8 that a small fraction of phytochrome, with properties different from the rest, can be found associated with certain cytoplasmic organelles.[361] The best evidence, however, has come from the study of phytochrome in seeds. It was mentioned on page 100, that in dry or freshly-imbibed seeds of several species, a small fraction of phytochrome, with very different properties from the rest, can be detected by sensitive *in vivo* spectrophotometers.[28, 30, 235, 408] The essential differences are that it is not subject to destruction, and that it is capable of the Pr to Pfr inverse dark reversion process. These two properties fit very closely the properties that would be predicted for the 'active' fraction of phytochrome postulated to account for the *Zea* and *Pisum* paradoxes.

In addition, there appear to be slight, but possibly significant, spectral differences between the fraction of phytochrome present in the freshly-imbibed seeds, and that found in high concentrations in dark-grown seedlings. Spruit and Mancinelli[408] using *Cucumis* and Kendrick *et al.*[235] using *Amaranthus* both published difference spectra for 'seed-phytochrome' and 'seedling-phytochrome'. In both cases the peak:height ratios were quite different for the two forms, but in *Cucumis* a shift of peak absorption of Pr in the seed-phytochrome, towards a longer wavelength, was noticed. The authors, however, in both papers, caution that the absorbance differences are very small and that the results should be regarded as preliminary.

It seems possible, therefore, that the phytochrome of the etiolated seedling may consist of at least two populations, one that is small in size and responsible for the metabolic and developmental responses to light, and the other, containing the bulk of the phytochrome, responsible for the *in vivo* spectrophotometric changes observable in the Ratiospect, but probably not involved in the mediation of the light responses. In a speculative mood, it could reasonably be suggested that 'bulk' phytochrome is an abnormal response to growth in darkness, since it may all be lost from the plant by the process of destruction within hours of receiving white light. Still speculating, it may be that 'active-phytochrome' is membrane-bound, while 'bulk-phytochrome' is present in the soluble, or mobile, phase of the cytoplasm.

Although these ideas seem reasonable, a word of warning is necessary. If the concept of bulk and active phytochrome is accepted as a means of reconciling the physiology with the spectrophotometry, it carries with it the corollary that all of the real correlations between *in vivo* spectrophotometric data, and growth data, are meaningless.

5.4 Hypotheses of phytochrome action.
I. Low-energy red/far-red reversible reactions

The rest of this chapter is devoted to an analysis of the several hypotheses proposed during the past decade, or so, to account for the way in which phytochrome regulates plant growth and development. In this section we deal with the ideas that have been advanced to explain the low-energy red/far-red photoreversible phenomena; the action of phytochrome under conditions of continuous irradiation is left until section 5.5. This is an artificial division since it would be reasonable to expect a single substance to operate through a single mechanism. Nevertheless, this is the way in which most of the hypotheses have been thought out.

Any plausible hypothesis of the mechanism of action must account for the two major aspects of photomorphogenesis. These are the amplification of the response and the inter-tissue specificity of the response. Furthermore, the hypothesis must be compatible with the known facts on the nature of phytochrome, and be able to explain both the long-term developmental effects of red light, and the very rapid effects on such phenomena as leaflet and chloroplast movement, described below. Even to say this is to beg the question to a degree, since there is an implicit assumption that there is a single primary reaction of Pfr, and that all the widely varying ultimate responses are manifestations of that primary reaction. It is quite feasible, as suggested by Mohr et al.,[301] that phytochrome is involved in qualitatively different primary reactions in different tissues and species, and thus the above assumption is not completely justified. This should not concern us too deeply at the moment, however, since it is obviously most sensible to follow the principle of Ockam and to investigate the simplest hypothesis first. Only if it becomes impossible to reconcile the known facts of phytochrome with the concept of a single primary reaction would it be fruitful to attempt the vastly more difficult task of elucidating a large number of different primary reactions.

(a) Pfr as an enzyme

The hypothesis that Pfr is an active form and Pr an inactive form of an enzyme was originally proposed by the Beltsville group on the basis of purely physiological experiments.[178] Considerable supporting evidence has been adduced from the study of the chemical properties of phytochrome – namely, phytochrome is a protein, and Pfr is much more reactive than Pr (see chapter 4).

The enzyme hypothesis satisfactorily explains the amplification of the initial stimulus, but it is more difficult to account for the specificity of action. It is necessary to postulate that the unknown reaction catalysed by Pfr is one that is fundamental to metabolism and capable of regulating the rates of many different metabolic pathways.

It is implicit in this hypothesis that the specificity resides in the particular metabolic, and differentiative, state of the cells stimulated by light, such that different cells react in different ways to the fundamental reaction catalysed by Pfr.

Rapidity of action can easily be explained with the Pfr as enzyme hypothesis since, if reasonable amounts of the hypothetical substrates were available, action could commence as soon as Pfr was formed from Pr. However, it has not been easy to obtain evidence supporting the hypothesis, since there are no clues as to in which area of metabolism the reaction might occur. The only serious suggestion made so far was based on minimal evidence and is not generally considered important at the present time. The idea came from the finding that the production of anthocyanin in disks of apple skin, floated on nutrient solution, is dependent on light. When the disks are kept in darkness, the formation of the volatile low molecular weight esters, characteristic of apples, was detectable by odour – in the light, no such odour was detectable. Since the esters are biosynthetically derived from acetate, these results led the investigators to conclude that anthocyanin synthesis was blocked in darkness at one of the initial steps, i.e., at the conversion of acetate to acetyl-Coenzyme A.[390] It was therefore suggested that Pfr is the enzyme which catalyses this 'activation' of acetate. This would be a suitable basic step for phytochrome action since acetyl-Coenzyme A is one of the most important intermediates in metabolism, being at the centre of many biosynthetic and catabolic routes.

It now seems likely, however, that the regulation of acetyl-Coenzyme A formation is mediated by phytochrome in some indirect way, since other investigators, using more reliable methods, have located other, later, steps in anthocyanin biosynthesis, also regulated by phytochrome. Furthermore, the effects of Pfr on anthocyanin synthesis are generally characterized by a lag phase between irradiation and pigment production, which would not be predicted for the direct action of Pfr on a single controlling step in the pathway.

(b) Phytochrome as a regulator of gene expression

An hypothesis which satisfactorily explains both the amplification and specificity of the developmental responses to phytochrome photoactivation was postulated by Professor H. Mohr of the University of Freiburg, in 1966.[295] This hypothesis, which has recently been comprehensively reviewed by its originator,[297] seeks to explain phytochrome action through the activation, or repression, of specific genes. The subsequent changes in enzyme levels are held to be responsible for the amplification, and the specificity is determined by the differentiational status of the genome of the particular cell.

Mohr suggests that the genome of each cell, at any point in development, consists of four classes of genes as characterized by their state of activity, and their potential for reacting with Pfr. The four classes are:

1. inactive genes,
2. active genes,
3. inactive genes which are activated by Pfr, and
4. active genes which are repressed by Pfr (Fig. 5.17).

Genes in category (1) are not required for the particular cell either in the presence, or in the absence, of Pfr – a suitable example would be a gene specifically related to root development being inactive in a shoot cell. Category (2) genes are required for

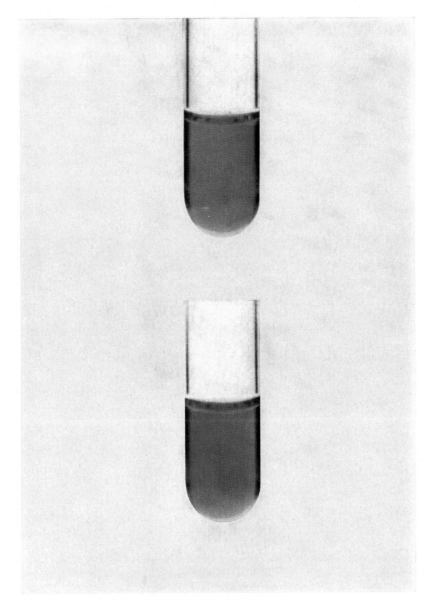

Plate 1. Photographs of highly purified large molecular weight rye phytochrome in highly concentrated aqueous solution (ca. 8 mg/ml). In the upper tube, the phytochrome is in the Pr form and thus appears blue; in the lower picture, the same tube has been irradiated with red light to convert Pr to Pfr which appears blue-green. (Photograph kindly supplied by G. Gardner of a preparation made with W. R. Briggs.)

the cell irrespective of Pfr availability, obvious examples being the genes coding for the synthesis of the enzymes of basic metabolism. The other two categories of genes are assumed to have a mechanism through which their activity can be regulated by Pfr, either directly, or indirectly. The particular genes which can react in this way to Pfr differ between differentiational states, and between species, thus providing for the specificity of the response. In this hypothesis, therefore, the specificity is not a property of phytochrome, but is determined by some mechanism which affects the properties of the genome and which is at present unspecified in molecular terms.

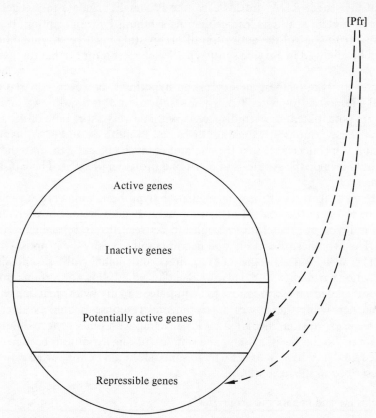

[Pfr]

Active genes

Inactive genes

Potentially active genes

Repressible genes

Fig. 5.17 Diagrammatic representation of the gene expression theory of phytochrome action (after Mohr).[296]

This is the first major point of conceptual difficulty in the hypothesis. If the regulation of gene activity is considered to be mediated by the mechanisms similar to those occurring in bacteria, it seems necessary to postulate that changes in the chemical properties of specific repressor molecules occur in relation to the differentiational status of the cells. These changes are necessary to account for the proposed onto-genetic differences in the reactivity of specific genes to Pfr. This difficulty could perhaps be lessened when, and if, specific proposals for the interaction between Pfr and the genome are put forward; at present it seems a formidable obstacle.

Mohr and his colleagues have accumulated considerable physiological and bio-chemical evidence, which they claim in support of the hypothesis. Most of it is based on the inhibition of Pfr-mediated responses by substances known to inhibit nucleic acid and protein synthesis. In particular, it has been shown that Pfr-stimulated anthocyanin synthesis in white mustard hypocotyls is inhibited by actinomycin-D, an inhibitor of DNA-dependent RNA synthesis,[256] and by cycloheximide, an inhibitor of protein synthesis.[306, 380] Actinomycin-D must be given before, or only shortly after, the start of irradiation for inhibition to be manifested, while cycloheximide is operative over a longer time period. While these results are indeed consistent with Pfr causing the synthesis of new messenger-RNA and new enzymes, they could probably be better explained in terms of Pfr action requiring the continued presence of messenger-RNA. In this case, the Pfr might regulate messenger-RNA translation in protein synthesis, rather than its synthesis by transcription. It is even possible for Pfr to control the activation of pre-existing inactive enzyme molecules rather than be involved in enzyme synthesis. (See chapter 8 for further discussion of this point.)

The major criticism of the gene expression hypothesis, however, is based on the timing of the responses to phytochrome photoactivation. It is now known that many responses are potentiated, or even displayed, within a very short time of the start of irradiation. These responses, which are discussed in more detail below, happen so rapidly that it is inconceivable that the chain of events connecting them to the initial stimulus could include the synthesis of RNA and protein molecules. There just is not enough time.

There are many features of the gene regulation hypothesis which are attractive for the function of Pfr in the control of development. It does seem, however, that the concept of a direct, or even close, relationship between Pfr and gene activity is too simplistic. The hypothesis has been modified in recent years by the proposal that the interaction between Pfr and the genome is indirect, and might involve several different primary reaction partners for Pfr. This modification would allow for certain responses, which do not involve gene expression, to be initiated rapidly by Pfr reacting with one reaction partner, while the slower developmental responses would be due to the control of gene expression, mediated by Pfr reacting with one or more different reaction partners. On this basis, a useful and simple working hypothesis becomes highly complicated, and it is difficult to see why this model should be supported rather than the one described in the next section.

(c) Phytochrome and membrane properties

As mentioned in the previous section, there are several phenomena controlled by phytochrome that occur with great rapidity. The classical example is the rotation of the chloroplast in *Mougeotia* which occurs with a time lag of about 2 minutes. Another example is the regulation of leaflet closing in *Mimosa pudicans*, the so-called sensitive plant, and several related species. In these plants, the leaflets close together on the transition from light to darkness. The closing movement is detectable within a minute, or so, of the removal of the light and the closing is complete within 30 to 40 minutes. This phenomenon, known as nyctinasty or sleep-movement, is under phytochrome control, as was first shown by Fondeville et al.[122] in 1966. They found that 730 nm light, supplied at the end of the light period, would inhibit the closing movement, and that the inhibition was detectable within 5 minutes of irradiation. Similar responses have since been seen in several other species exhibiting nyctinastic leaf movements.[195]

In one closely related member of the Leguminosae, *Albizzia julibrissin*, it is known that the closing movement is associated with a rapid loss of electrolytes from the regions responsible for the movement.[213] At the base of each leaflet, and of each leaf, there are localized concentrations of large cells forming an organ known as a *pulvinus*. The closure of the leaflets occurs when the turgor pressure of these cells is rapidly, and massively, decreased, probably by electrolytes being pumped, or leaked out of them into the nearby vascular tissues. Using an electron probe device, Galston and coworkers have been able to show that red and far-red light bring about rapid movements of potassium between the pulvinal cells and the neighbouring cells in *Albizzia*.[371] (These phenomena are considered in more detail in chapter 8.)

A yet more remarkable phenomenon, observable within seconds, is the photoreversible adhesion of root tips to a glass surface.[415,416] When secondary root tips of etiolated barley, or mung bean seedlings, are excised and placed in a negatively charged glass dish they float on the surface of the liquid, or sink to the bottom of the vessel. After irradiation with red light, they immediately stick, by their tips, to the surface of the glass. Subsequent irradiation with far-red light will immediately release the roots from the glass. It has been suggested that the basis of the adhesivity is the electrical potential existing between the outer surface and the interior of the root, and that this potential difference is rapidly changed by phytochrome photoactivation. This is indicated because the roots will only stick to a glass surface that has been negatively charged. The requirements of the adhesion phenomenon are many and complex and are described in detail in chapter 8.

Direct evidence for the involvement of electrical potential changes in this phenomenon has been obtained by the measurement of potentials in mung bean roots by Jaffe.[210] This investigation has shown quite clearly that the potentials across the root change in response to red light, and that the change is substantially reversible by far-red light. These results have since been confirmed by Newman and Briggs[315] who have measured a 15 second lag between red light and potential difference changes. (See also page 161.)

Red light treatment of certain cereal leaves causes a very rapid increase in gibberellin level which peaks at about 10–15 minutes after the start of the treatment.[12,352] This may be considered as a rapid biochemical effect of phytochrome activation, and in fact recent work has shown that increased gibberellin levels are produced in leaf homogenates irradiated with red light.[353] Although enzyme action is probably necessary for this response, it is highly unlikely that changes in the amounts of the relevant enzymes could occur in the short times available. These various rapid displays of phytochrome action are discussed in more detail in chapter 8, and have been described here merely to demonstrate that any theory of phytochrome action must take into account its rapidity of action.

Other evidence for rapid action of phytochrome comes from phenomena which are not themselves displayed rapidly, but which, nevertheless, can be shown to have been potentiated within a few minutes of the brief red light treatment. One example is in the photoperiodic control of flowering in *Pharbitis nil*[125] and *Kalanchoe blossfeldiana*,[294] in which far-red reversibility of a red light break in the middle of a long night is lost within 5 minutes of the red light. The flowering responses themselves, however, do not become measurable until weeks later.

Another similar type of response occurs in the red light potentiation of the germination of Grand Rapids lettuce seeds. These dark-dormant seeds can be caused to germinate either by brief red light treatment or by incubation in gibberellic acid.

Normally the red light treatment is fully reversible by far-red light given immediately afterwards. If, however, the seeds are placed immediately after red light in a sub-threshold concentration of gibberellic acid (i.e., a concentration so low that it will not itself cause germination) then subsequent far-red light is no longer able to suppress the germination.[17] Germination, of course, does not begin until many hours later. In both of these cases, then, phytochrome appears to have had its action within a very short time of its photoconversion; the lag between this action and the ultimate developmental processes must be due to other limiting steps along the way.

There is, therefore, irrefutable evidence that phytochrome can act very rapidly, almost instantaneously in fact, either to potentiate a slow developmental event, or to cause immediate displays such as leaflet movement, chloroplast rotation, root adhesion and electrical potential changes. These observations cannot be accounted for by a hypothesis involving the mediation of nucleic acid, and protein synthesis, in the action of phytochrome. They can, however, be reconciled if it is assumed that the photoactivation of phytochrome somehow leads to immediate changes in the permeability, or other properties, of certain critical cellular membranes. All the rapid displays described above, such as leaflet closure, chloroplast rotation, root adhesion, and potential changes, would appear to involve mechanisms centring on membranes, in one way or another.

The hypothesis has thus been proposed by Borthwick and Hendricks[182] that phytochrome is a component of, or acts directly on, certain important cellular membranes such that the permeability of these membranes is regulated by light treatment. In this way, substrates, enzymes, hormones, etc., may be released from cellular compartments and allowed to interact, thus bringing about the more typical, long-term developmental responses to light treatment. The specificity of the phytochrome responses is once more thought to reside not in the primary reaction of phytochrome, but rather in the genetic status of the individual cells.

We may fruitfully consider the possible mechanisms through which the photo-activation of phytochrome might lead to such marked permeability changes. The most likely possibility is that phytochrome is indeed a component of the membrane, as suggested by the polarized light experiments of Haupt and others (see section 5.1). As pointed out in chapter 4 it is known that the changes from Pr to Pfr and *vice versa* involve a conformational change in the protein. Furthermore, through ingenious utilization of polarized microbeams, Haupt et al.[173] have shown that the axis of orientation of the phytochrome molecules in the plasmalemma of *Mougeotia* is

Fig. 5.18 Demonstration of the change in orientation of phytochrome molecules in *Mougeotia* upon photo-conversions:

(a) Part of a *Mougeotia* cell in cross-section (above) and in surface view (below). (i), (ii) and (iii) before, during and after irradiation with the microbeam, respectively, which is shown in (ii). Chloroplast and orientation of phytochrome molecules are shown in (i) and (iii).

(b) As in (a) but with phytochrome molecules omitted. Double arrows = electrical vector of polarized red (R) and far-red (FR) light respectively.

Note that the curvature of the chloroplast is used as a diagnostic test for the presence of Pfr. Photoconversion of Pr and Pfr requires red light polarized parallel to the cell axis; reversal requires far-red light polarized at right angles to the cell axis; therefore the chromophores must reorientate through 90° upon each photoconversion (from Haupt et al.).[174]

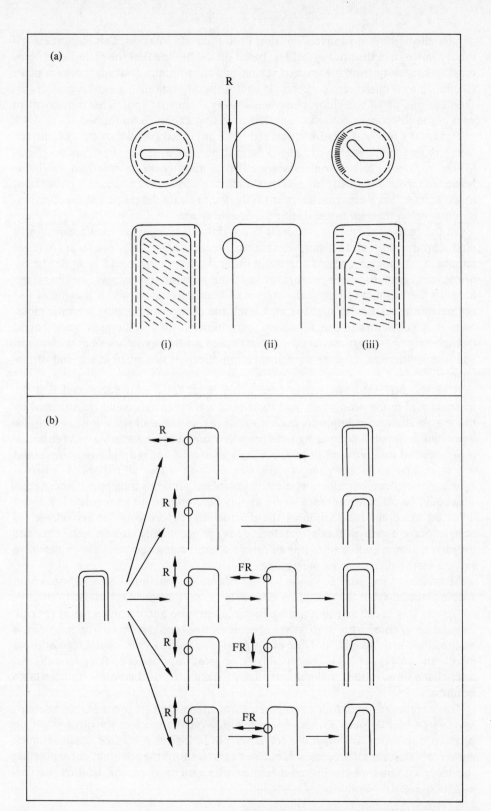

rotated through 90° on conversion from Pr to Pfr and *vice versa*. This elegant experiment, shown in outline in Fig. 5.18, is based on the finding that the plate-like chloroplast moves away from a localized region of Pfr molecules towards a region of Pr (see Fig. 5.6). Thus, if a microbeam of light is used to establish a local region of Pfr, then the margin of the chloroplast twists away from that spot. This movement of part of the chloroplast is used as an indication that Pfr has been formed.

The results in Fig. 5.18 show that red light polarized parallel to the cell surface converts Pr to Pfr, while red light polarized normal to the surface has no effect. Moreover, once Pfr has been established by a microbeam of polarized red light, polarized far-red light will only reverse the effect if its plane of polarization is normal to the surface. Such experiments beautifully demonstrate the reorientation of phytochrome within the membrane during photoconversion.

On the basis of these discoveries, it is not difficult to imagine localized arrays of phytochrome molecules causing very big changes in the molecular properties of those regions of the membrane. At the most simplistic level, one could imagine the reorientation, and change in conformation of the phytochrome arrays, yielding large holes in the membrane rendering it permeable to substances which previously did not penetrate. At a more sophisticated level, the changes in the phytochrome molecules may be factors in the activation, or inhibition, of specific membrane-bound transporting agents or 'carriers'. A model based on this hypothesis but with modifications to account for long-term irradiation effects is described at the end of this chapter.

A rather different attitude has been taken by Jaffe[211] who discovered that the adhesivity of mung bean roots, and their electrical potentials could be regulated by the mammalian neurohumour, acetylcholine. From these observations (which are described in detail in chapter 8) Jaffe proposed the following scheme: acetylcholine is synthesized and released in the root cells as a result of red light treatment, and moves at random to many target sites. One of these is the cell membrane where it mediates the effects on adhesion, electric potentials, and ion transport. Other target sites could be other membranes where metabolic events could be regulated. Far-red light, on the other hand, induces the destruction, or prevents further release, of acetylcholine. This hypothesis, therefore, does not necessarily imply a membrane site for phytochrome itself; rather it implicates a second messenger which itself can have several varied effects on cell membranes.

At present, very little is known of the degree of functional compartmentation within plant cells, but the available evidence indicates that such separation of reactive components is of extreme importance in the integration and organization of the cell. Regulation of compartmentation in a specific manner could have far-reaching effects on the ultimate developmental fate of the particular cell. In addition the regulation of membrane properties could conceivably be of great importance in the inter-cell, and inter-tissue, integration of developmental processes, as mediated by transportable hormones.

The membrane-regulation hypothesis, then, irrespective of the molecular details, appears to be the most attractive current idea on the mechanism through which phytochrome controls both the very rapid physiological processes, and the much slower, developmental processes. It also appears to afford the possibility of explaining the more complex phytochrome-mediated phenomena requiring continuous illumination, as is shown in the next section.

5.5 Hypotheses of phytochrome action.
II. The high-energy reaction

The elegant experiments carried out by Hartmann[167] and described in chapter 3 prove to most reasonable minds that phytochrome is the photoreceptor for the high-energy reaction, at least for the spectral regions above 550 nm. It is necessary therefore, for us to consider the possible mechanisms whereby phytochrome mediates these effects. The essential points to be explained are: that maximum action is obtained at specific and critical Pfr/Ptotal values; that maximum action depends on prolonged irradiation for considerable time periods, and that the physiological responses are normally irradiance-dependent over a wide range.

(a) The destruction hypothesis of phytochrome action in continuous light

Two conspicuous things happen to phytochrome when an etiolated plant is irradiated for a long period (i.e., several hours). Firstly, after a very short period, a dynamic equilibrium between Pr and Pfr is set up and subsequently, although interconversions continue to occur, the proportion of Pr and Pfr remain constant, increased irradiance only serving to increase the rate of photoconversion in both directions. Secondly, in wavelengths which set up a substantial proportion of the total phytochrome as Pfr the destruction process rapidly removes the vast bulk of the total phytochrome. These considerations have been paramount in attempts to construct a hypothesis for phytochrome action under continuous irradiation.

Hartmann,[167] for example, has constructed a hypothesis based on the concept that maximum action will occur under those conditions that allow Pfr, even at very low concentrations, to be present for a very long period of time. It is thus a question of balancing the need to have a sufficiently high level of Pfr concentration to maintain action, and the opposing need to keep the Pfr concentration as low as possible in order to keep destruction to a minimum. Hartmann's hypothesis, based on his work with lettuce seedlings satisfactorily accounts for maximum action being obtained at specific Pfr/Ptotal values, and being dependent on prolonged irradiation, but does not easily account for the irradiance dependence of the high-energy reaction. On this model, which is shown in diagram form in Fig. 5.19 an equilibrium is set up at a particular Pfr/Ptotal value, at which destruction is kept to a minimum, yet which still allows the maintenance of sufficient Pfr to function for as long a time as possible. The function of Pfr on this model is to combine non-photochemically with an unknown metabolite (X) in which state it is capable of initiating the processes leading to the developmental changes. If Pfr is considered to be resistant to destruction while it is in combination with X, then it can be seen that the Pfr $+$ X $\rightarrow$ PfrX reaction will be in competition with the destruction reaction. Since destruction is only dependent on the steady-state concentration of Pfr while the binding to X is dependent on the concentrations of Pfr, X, and PfrX, it is clear that the optimum photostationary state will be related to the steady-state concentration of X. This point allows for the existence of different optimum photostationary states in different tissues of the same organism.

This hypothesis assigns a critically important role to phytochrome destruction, and on these grounds it has been subject to criticism. In particular, Hillman[193] has pointed out that the assumption that all, or at least a constant proportion, of the phytochrome of an etiolated plant is active is based on scanty evidence. On the other hand, powerful evidence that Pfr destruction can be of great importance in one high-energy reaction comes from the results of Grill and Vince.[156] In these experiments on anthocyanin synthesis in turnip seedlings, pretreatment with blue or red light decreased the subsequent response of the tissues to a long period of exposure to far-red light. This effect could be observed with short periods of red light followed by 2 hours of darkness prior to the continuous far-red light. Furthermore, the inhibition caused by the red light pretreatment could be reversed by an immediate subsequent short period of irradiation with far-red light. It was therefore assumed that the effect of the red light was to establish a high proportion of Pfr which was then rapidly

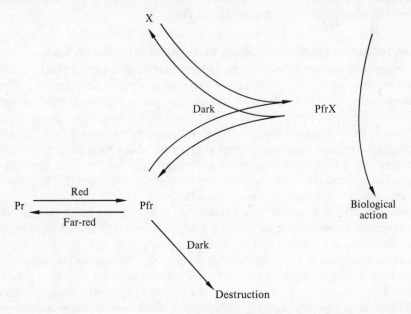

Fig. 5.19 The Hartmann hypothesis for the action of phytochrome in the high-energy reaction (see text for explanation) (from Hartmann).[167]

destroyed in the subsequent darkness, so that insufficient total phytochrome for maximum action was left by the time the far-red light was given. Strong supporting evidence for this conclusion came from the finding that the inhibitory effect of a brief pretreatment with red light could be virtually eliminated by holding the tissues at low temperatures, or in an atmosphere of nitrogen, for the intervening dark period. Since the destruction process is a metabolic, oxygen-dependent reaction, both the low temperature treatment, and the absence of oxygen, would tend to inhibit it, and thus to preserve total phytochrome during the dark period. It must be stressed, however, that although this is compelling evidence for the importance of destruction in this particular reaction, by no means all high-energy reactions exhibit the same properties.

With reservations, then, Hartmann's hypothesis accounts for the far-red peaks of action in the high-energy reaction, on the basis of a competition between Pfr action and Pfr destruction. It is more difficult, however, to explain the observed irradiance dependence (Fig. 3.6) of the high-energy reaction on the basis of this hypothesis. The critical feature of the hypothesis is the maintenance of a specific photostationary state, which can be achieved at relatively low energies of monochromatic irradiation.

Although increased irradiance will increase the rates of interconversion of Pr and Pfr, the photostationary state will not be changed, and thus there appears to be no obvious basis upon which the irradiance dependence can be explained. This is probably the major drawback of this hypothesis and has been responsible for its failure to achieve unanimous support.

Another problem has been the fact that several high-energy processes seem to operate most efficiently at wavelengths at which the Pfr concentration is kept high, and thus at which the destruction rate should also be high.[38] Furthermore, other processes show irradiance dependence with continuous far-red light, even when given steady background radiation in the red regions, maintaining high levels of Pfr.[38] These latter complications are very difficult to interpret on any hypothesis of phytochrome action, and may only be reconciled with the reluctant assumption that a further far-red absorbing photoreceptor is involved. Hendricks et al.[38] suggest, in fact, that PfrX in Hartmann's model (Fig. 5.19) might have different absorption properties from Pfr and could thus be a candidate for a further far-red absorbing photoreceptor. There is no real evidence for this suggestion, but since we are by now only speculating, this idea is as good as any other.

It should be pointed out that Hartmann has striven to modify his model to take account of the irradiance dependence of the high-energy reaction. The idea is that newly-formed Pfr is somehow different from older Pfr and thus, as the irradiance increases and the rate of cycling between Pr and Pfr increases, the proportion of newly-formed Pfr increases. If this new Pfr is better able to react with X, then the subsequent developmental processes (Fig. 5.19) would be expected to be related to the quantum flux density received.[167] Whether this is the correct solution remains to be seen, but it should be borne in mind that the hypothesis, even in this modified form, includes a critically important role for the destruction process, with all that implies.

(b) Phytochrome as a membrane transport factor

In this hypothesis, which is represented schematically in Fig. 5.20, phytochrome is thought to be oriented in certain important cellular membranes, and that its primary function is to transfer a critical metabolite (X) from one side of the membrane to the other.[397] The membrane is thought to separate the site of synthesis, or storage, of X from the site of its utilization in reactions which initiate the morphogenic processes. The nature of X is not specified, although it is possible that it is an important hormone, perhaps even of a macromolecular nature, and that the reactions in which it takes part may depend on the particular differentiative status of the irradiated cell. To account for amplification, these reactions must, in all probability, include the regulation of the expression of specific genes, and the activation, or inhibition, of specific enzymes. Thus, in common with most other hypotheses of phytochrome action, this model assigns the specificity of the morphogenic response to the specific differentiative status of the target cells.

Up to this point, this hypothesis does not differ significantly from the membrane permeability hypothesis of Hendricks and Borthwick[182] discussed above. In all other hypotheses, however, it is held that Pfr is the active moiety. This model assumes that the transfer of X across the membrane is actually driven by the photoconversion of Pr to Pfr and *vice versa*. Thus, on this concept, the primary action of phytochrome is not a function of the chemical reactivity of Pfr, but occurs simultaneously with the production of Pfr from Pr. When Pfr is produced from Pr by a single saturating irradiation with red light, a certain quantity of X is transferred across the membrane where it interacts with its specific reaction mechanisms (A in Fig. 5.20). Photoreversibility by far-red irradiation would be due to the reverse transfer of X across the membrane, and rapid irreversible action of phytochrome would be caused by the immediate interaction of a part of X with A. This would explain the rapid action of phytochrome in developmental responses such as the interaction with gibberellin in lettuce seed germination described above.[17] It is also obvious that if the rate of production of A was slow, then a particular response would be reversible by far-red as long as significant amounts of both Pfr and X° were left. On the other hand, if A is plentifully available then X° would be rapidly used up, meaning that photoreversibility would only exist for a short time after red irradiation. Large variations in the time courses of the loss of photoreversibility exist in practice (Table 3.5).

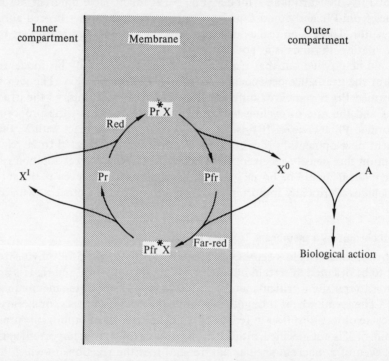

Fig. 5.20 The transport factor hypothesis for phytochrome action (see text for explanation) (from Smith).[397]

When the phytochrome molecules in the membrane are exposed to continuous light of wavelengths that are absorbed by both Pr and Pfr, then a dynamic photostationary state would be established in which X is shuttled backward and forward across the membrane. Under these circumstances, maximum action would be expected to occur at a specific photostationary state which maintains the maximum steady-state concentration of X outside the membrane. Since Pr is proposed to bind X on the inside of the compartment, and to release it upon photoconversion on the outside of the membrane, then maximum binding will be favoured when the steady-state concentration of Pr is kept high, i.e., when binding of X^I is preferred. The same argument applies to the binding of X to Pfr outside the compartment; thus minimum binding of Pfr to X° will occur when the steady-state concentration of Pfr is kept low. On this basis, maximum steady-state concentration of X° will be maintained with a low Pfr/Ptotal proportion. Thus, the far-red action maxima and the requirement for a low photostationary state can be explained without the involvement of destruction.

It is also possible to account for the irradiance dependence of the high-energy reaction since, as long as A is not rate limiting, increased radiant flux will increase the rate of cycling between Pr and Pfr, and thus increase the availability of X° for its reactions with A.

Phytochrome on this basis, therefore, is thought to be a specific carrier molecule, or permease, which brings about an internal change in the hormonal, or metabolic status, of the target cells. The ultimate physiological or morphogenic response is deemed to be due to this changed hormonal or metabolic status. It must be stressed that this model is purely speculative and is difficult to test. It would predict, however, that isolated phytochrome would be found to bind X, if only we knew what X was. Studies on membrane fractions containing phytochrome should be valuable since it is likely that solubilized phytochrome may have lost any binding properties specifically associated with its orientation in the membrane.

Whether this model is shown to be correct, partially correct, or totally inadequate is to an extent immaterial. It is merely intended to show that all of the various loose ends of phytochrome action can be collected together and tied up as long as certain simple assumptions are made. These are:

1. that active phytochrome molecules are membrane components;
2. active phytochrome has binding sites on the molecule for a specific metabolite; and
3. the binding sites are anisotropic, i.e., on different sides of the membrane, in the Pr and Pfr form.

Finally, a strong cautionary word is necessary. The precise and elegant experiments described in the preceding three chapters have provided invaluable evidence on the functional role of phytochrome, but we must continually remember that the plant has not become specifically adapted for growth under monochromatic light sources, or to respond to alternating sequences of red and far-red light. It is important at all times to consider the function of the photomorphogenic reaction systems in the natural environment. Phytochrome is not merely a photoreceptor for the red and far-red action maxima of the photomorphogenic reaction systems. It is, *par excellence*, a mechanism through which the plant can detect the spectral quality of the radiation it is receiving, and adjust its growth pattern in accordance. The photostationary state of phytochrome is determined by the spectral distribution of energy in the red and far-red wavelength regions, thus providing a highly sensitive mechanism whereby the plant is able to respond both metabolically, and developmentally, to the changing irradiation conditions under which it is growing. This point is returned to in chapter 7.

5.6 Conclusions

Phytochrome is now known to be concentrated in the growing regions of etiolated plants and, within cells, is probably located in various membranes, although the evidence is not certain on this latter point. The phototransformations appear to involve the same sequence of intermediates *in vivo* as *in vitro*, but there is evidence that under continuous high-irradiance levels, one of the intermediates accumulates in relatively high proportions. A number of dark transformations occur *in vivo* including dark reversion of Pfr to Pr, which is known only in certain dicotyledonous plants, destruction of Pfr to an unknown product (found in almost all plants), inverse dark reversion of Pr to Pfr which is restricted to certain seeds, and apparent resynthesis of Pr. Attempts to correlate the degree of photoconversion of Pr to Pfr with the developmental effects obtained, have been largely satisfactory, although two striking and inexplicable paradoxes remain. These can only be understood at present in terms of there being a small active fraction, and a large inactive (bulk) fraction, of phytochrome, which in certain tissues can exhibit different photoconversion characteristics.

Several hypotheses of phytochrome action have been proposed including Pfr as an enzyme, Pfr as a direct or indirect regulator of gene expression, and Pfr as a modifier of membrane permeability. For the high-energy reaction, it is necessary to account for the wavelength of maximum action, the long-term nature of the response and the irradiance dependency. Attempts to fulfil these requirements involve, on the one hand, the concept that newly-found Pfr is more reactive than pre-existing Pfr, and a competition between Pfr action and Pfr destruction exists; on the other hand, it has been proposed that phytochrome is present in a membrane where it acts as a permease in which the transport of a critical metabolite across the membrane is driven by the photoconversions.

A full evaluation of these various hypotheses cannot yet be made. The subject is, however, returned to in chapter 9 after the physiology and biochemistry of photomorphogenesis have been covered in more detail. The aim of this first half of the book has been to analyse and dissect what is known of the photoresponsive reaction mechanisms. The rest of the book is concerned with integrating this knowledge as far as possible into, hopefully, an understanding of the role of light in the vegetative development of the plant.

6

The photocontrol of seed germination

Seed germination was one of the first developmental responses to be recognized as being capable of photoregulation. References to effects of light on germination can be found, for example, in the writings of Ingenhousz and Senebier in the late eighteenth century.[25] The first recorded experimental demonstration, however, was not published until 1860, with the observation by Caspary[68] that the seeds of *Bulliarda aquatica* (*Tillaea aquatica*) would not germinate in darkness. By the turn of the century, the existence of 'photoblastism' – as the photoregulation of seed germination ultimately came to be called[109] – was firmly established and large lists of species whose germination was either promoted, or inhibited by light could be compiled. For example, Kinzel[236] in 1908 recorded over 600 species which were controlled in this way. The final chapter in the history of photoblastism began in the mid thirties with the observations by Flint and McAllister[120, 121] of the opposite effects of red and far-red light on the germination of lettuce seeds. Although this work was not followed up immediately, it formed the basis of the discovery of the red/far-red photoreversibility of germination, and ultimately of phytochrome itself, by Borthwick and Hendricks and their colleagues at Beltsville, as is described in chapter 3.

6.1 Positive and negative photoblastism

White light can affect germination in one of two ways – it can stimulate germination, or it can inhibit germination. In addition, it often has no effect whatsoever. When germination is stimulated by *white* light, the seeds are commonly said to be positively photoblastic, but if *white* light inhibits germination, then the seeds are negatively photoblastic. Note the insistence on white light in the above definitions. It will be seen below that the stimulatory and inhibitory actions of white light are caused by different spectral regions, and that the germination response of a specific seed is determined by its relative sensitivity to these two wavelengths. Thus, although positively photoblastic seeds can be caused to germinate by white light, this stimulation can often be reversed by subsequent treatment with light of a restricted spectral region, namely far-red light, while inhibition of the germination of a negatively photoblastic seed by white light can similarly, in many cases, be reversed by treatment with red light. The present view is that positive and negative photoblastism are simply two facets of a single phenomenon, and that only one photoreceptor is responsible for the two processes. The evidence for this view is discussed later.

The classical requirements for the germination of non-dormant seeds are simply favourable conditions of moisture, temperature and aeration. Light is not included and in fact the majority of seeds are not affected, either positively or negatively, by light. Such seeds are often said to be non-photoblastic, but it is interesting that many non-photoblastic seeds can be made light-requiring by external conditions of stress.[110] The conditions which can impose light sensitivity include high osmotic pressures,[219,376] and the application of certain inhibitory substances.[318] There are cases, e.g., radish, in which high osmotic potentials cause a normally insensitive seed to become light inhibited.[271] Experiments on imposed photoblastism seem to indicate that the processes of germination in photoblastic and non-photoblastic seeds are similar, except that in the one case they can be switched on or off by light of the requisite wavelength.

As mentioned above, a very large number of species are known whose germination is regulated by light. It is not the purpose of this book to present such encyclopaedic lists, but Table 6.1 gives a small selection of species whose germination behaviour is

Table 6.1 Selected species with seeds showing light sensitivity (from Wareing[451] with additions).

Light-promoted	Light-inhibited
Adonis vernalis	*Ailanthus glandulosa*
Bellis perennis	*Amaranthus caudatus*
Chenopodium album	*Cucumis sativus* (varieties)
Digitalis purpurea	*Euonymus japonica*
Epilobium hirsutum	*Forsythia suspensa*
Erodium cicutarium	*Hedera helix*
Fagus sylvatica	*Nemophila insignis*
Helianthum chamaecistus	*Nigella damascena*
Iris pseudaccorus	*Lamium amplexicaule*
Juncus tenuis	*Lycopersicon esculentum* (varieties)
Lactuca sativa (varieties)	*Phacelia tanacetifolia*
Lactuca scariola	*Phlox drummondii*
Lepidium virginicum	*Tamus communis*
Lythrum salicaria	
Oenothera biennis	
Ranunculus scleratus	
Rumex crispus	

well established. It is interesting to note how few cultivated species are included, and it is common knowledge that the germination of the seeds of cultivated plants is generally unaffected by the light regime. This is presumably due to artificial selection for homogeneous germination practised by hundreds of generations of crop-breeders. Such homogeneity of germination behaviour is of great advantage in experimental work, but it is conspicuously lacking in the seeds of weed species, the germination of which is generally more interesting from the point of view of photomorphogenesis. Weed species show a considerable degree of polymorphism in germination behaviour manifesting itself in great variability in germination rate. In certain cases, indeed, individual plants can produce seed some of which requires light for germination while the rest does not. Polymorphism of this type is clearly of great adaptive value since it results in intermittent germination of seed over long time intervals, ensuring that at least some seeds of a population will germinate while conditions are conducive to successful seedling establishment.

The amount of light energy required for germination varies from plant to plant. The most sensitive plants need only very small amounts of light energy (e.g., *Lactuca sativa* and *Lythrum salicaria*) while other seeds can require up to 24–48 hours continuous illumination (e.g., *Epilobium hirsutum*). Attempts have been made in the past to classify seeds according to their quantitative light requirements. For example, Isikawa and Shimogawara,[209] in 1954, defined four categories of positive photoblastism as follows:

1. The *Nicotiana* type requiring only a single short exposure of *ca*. 1 minute to irradiances of *ca*. 1000 lux.*
2. The *Plantago* type requiring a single exposure of somewhat longer time (*ca*. 1 hour) at similar irradiances.
3. The *Epilobium* type requiring a prolonged light exposure of *ca*. 24 hours.
4. The *Hypericum* type requiring repeated daily exposures of several hours per day.

These distinctions are obviously artificial and later work has shown them to be of doubtful value. In many cases of seeds or spores whose germination appears to require long periods of illumination, the stimulatory effect of the long light period can be replaced by two short periods of illumination separated by several hours of darkness. For example, in the light-stimulated germination of *Dryopteris crassirhizoma* spores, as high a germination percentage can be obtained with two, 30 minute periods of irradiation separated by 15 hours darkness, as with 16 hours of continuous irradiation.[208] Similar results were reported by Borthwick[33] for seeds of *Puya* and *Pinus* species.

It appears that, in reality, there are only two types of positive light requirements for germination, distinguishable on the basis of energy, and time of irradiation requirements. In the first type, exemplified by lettuce (*Lactuca sativa*) and probably including the first two categories of Isikawa and Shimogawara,[209] the percentage germination is saturated at low total energies of irradiation, and below saturation level the reciprocity law holds true. In the other type, saturation of germination percentage is not energy dependent but is related to the time period over which irradiation occurs. In this type of germination, intermittent irradiation is often as effective as continuous irradiation. The explanation of these two different response patterns will become evident later.

6.2 Photoperiodism in seed germination

The fourth category of Isikawa and Shimogawara[209] implies a requirement for periodic irradiation regimes for the germination of some seeds. Indeed, Black and Wareing[26] in 1954, showed that under certain conditions, the germination of birch (*Betula pubescens*) seed was photoperiodic in nature. Although unchilled seeds at 20°C, would germinate after one irradiation treatment, at 15°C at least eight consecutive daily periods of light treatment were required. For full germination, moreover, photoperiods of about 20–24 hours were required, shorter photoperiods giving

---

* The lux is a photometric unit of irradiance in which the spectral sensitivity of the detecting instrument is chosen so as to reflect as closely as possible the spectral sensitivity of the human eye; photometric units cannot easily be converted to radiometric units in which all radiant energy is measured, and thus their usage is not recommended, although they were commonly employed in earlier work (see chapter 2).

progressively lower germination percentages. This type of response is not identical with the photoperiodic regulation of flowering,[343] since a critical dark period cannot be defined; the relationship between the length of the dark period and germination response is simply quantitative.

The germination of certain other seeds has been reported to be a short-day phenomenon in that extension of the daily period of illumination beyond a certain period leads to a reduction in final germination percentage. For example, the germination of *Epilobium cephalostigma* seeds is greater under irradiation periods of between 10 minutes to 21 hours, than it is under continuous illumination.[206] Similar observations have been made on seeds of *Tsuga canadensis*.[321] Again, it appears that these short-day responses are different in nature to the photoperiodic regulation of flowering, although critical analysis of the phenomenon by the light break method has not been carried out.

6.3 Effectiveness of light in relation to imbibition time

In most light-sensitive seeds, whether positive or negative in their light responses, the effect of a specific light treatment varies considerably with time of imbibition in darkness before application of the light treatment. Figure 6.1a shows the response of lettuce *var*. Grand Rapids seed to a short period of white light, given at different times after the start of imbibition, as observed by Evenari and Neumann.[112] It is clear that a peak of photosensitivity occurs after approximately 8 hours imbibition, followed by a steady decline in sensitivity. Ultimately, photosensitivity is completely lost. The relationship between times of imbibition and photosensitivity are, of course, different for different seeds. Even using the same variety of lettuce seeds, other workers have observed different curves for the development of photosensitivity. Figure 6.1b shows the results of Ikuma and Thimann[204] using red light and contrasts markedly with those of Evenari and Neumann.[112] These different responses may be due to differences in the conditions under which the two batches of seed matured on the mother plants or to differences in storage conditions. Another possibility is that white and red light act slightly differently, but this interesting point has not really been taken up.

Other species, naturally, show much more marked differences in the time-course of development of photosensitivity. An extreme case is *Pinus taeda*, which develops increasing photosensitivity over a period of more than 60 days.[437] In other cases, e.g., *Oryzopsis miliacea*, photosensitivity is achieved rapidly, but is not lost, seeds retaining their response to light for up to 28 days.[249] It is obvious, then, that large intrinsic differences exist between different species in their sensitivity to light.

It is even more interesting to consider the responses of some white light-inhibited seeds to irradiation given at varying times after the start of imbibition. In *Nigella*[207] and *Phleum pratense*[273] light given during the first 30 hours of imbibition leads to increased germination over the dark period, whereas light given after this period leads to inhibition of germination. In both species, seed ultimately (in 3–5 days) loses all sensitivity to light. In the case of *Nigella*, the pattern is even more complex since, in the initial 30 hour period, although short light treatments, of up to 3 hours, stimulate germination, longer periods (12–24 hours light) give decreasing stimulation and even considerable inhibition. Thus, the sensitivity to light in these seeds is a very complex function of duration of imbibition and duration of light treatment. As

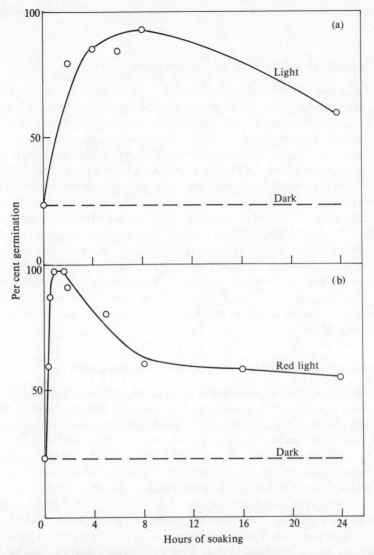

Fig. 6.1 Photosensitivity of lettuce *var*. Grand Rapids seeds in relation to time of imbibition.
(a) Data of Evenari and Neumann.[112]
(b) Data of Ikuma and Thimann.[204]
Note the same general pattern in both experiments, but the different time-courses.

implied above, and discussed in more detail in section 6.7, the promotive and inhibitory actions of white light are due to different spectral regions, and it may thus be possible to explain the complex patterns exhibited by *Nigella*, and similar seeds, on the basis of different sensitivities to these spectral regions at different times of imbibition.

It is commonly assumed that dry seeds are not photosensitive. However, Evenari and Neumann[111] in 1953 showed that if dry lettuce seed is maintained at relatively high humidity in light or darkness, the final dark germination of those seeds, stored in the light, is significantly higher than those kept in darkness. Such considerations are worthy of note when conducting experiments on the photoregulation of seed germination.

6.4   Photodormancy and skotodormancy

As Fig. 6.1 shows, if imbibed lettuce seed is kept in darkness for more than *ca*. 18 hours, the seeds begin to lose their sensitivity to light. If this process is continued, the seeds ultimately become wholly insensitive to light and within 3–4 days are completely dormant. This dormancy, which can be induced in many normally light-requiring seeds, by maintenance in darkness for long periods of time, is known as 'skotodormancy' (*skoto* ≡ dark). An analogous phenomenon is the induction of 'photodormancy' in many light-inhibited seeds, by prolonged exposure to germination-inhibiting light sources. The depth of photodormancy appears to be related to the irradiance of the light used, whereas the depth of skotodormancy is dependent on the length of the dark period.[110] Both processes appear to have similar characteristics since dormancy ensues following prolonged exposure to conditions which prevent germination (i.e., darkness for light-requiring seeds and continuous light for light-inhibited seeds). Both photodormancy and skotodormancy can normally be broken by cold treatment. Although the development of these forms of dormancy was a mystery for a long time, it is now possible to postulate mechanisms to account for them (see page 137).

6.5   Interaction between light and temperature

The germination of photosensitive seeds in darkness, and under light treatments, can be drastically modified by temperature. The range of effects of temperature, and temperature changes, is enormous and it would not be productive to attempt full coverage here. It is, however, constructive to consider the situation in lettuce seeds, perhaps the favourite photomorphogenic material of the last three decades.

Lettuce *var*. 'Grand Rapids' is a classic light-requiring seed, but its photosensitivity varies markedly from batch to batch. Some batches, indeed, germinate in total darkness at normal experimental temperatures of 20°C to 25°C. At higher temperatures dark germination falls off, and the seeds become light-requiring. At even higher temperatures, germination is prevented completely, both in dark and light. (If seeds are maintained in the imbibed state at these high temperatures of *ca*. 30–35°C for several days, either in light or darkness, they become 'thermodormant' – i.e., they lose the capacity to germinate at normally favourable temperatures irrespective of the light conditions.)

The pattern is, then, that at high temperatures no germination occurs whatever the conditions, at intermediate temperatures the seeds become light-requiring, while at lower temperatures, they germinate in darkness. This is illustrated for five different batches of Grand Rapids lettuce seed in Fig. 6.2.

The wide range of temperatures at which the different batches become light-sensitive is very clear in this figure. If it is assumed that seeds which germinate in darkness have overcome, or bypassed, the block to germination represented by the photocontrol mechanism, then the lower the temperature, the greater the number of seeds which have the ability to overcome, or bypass, the block. That the photo-mechanism is still present at lower temperatures can be shown by placing the seeds

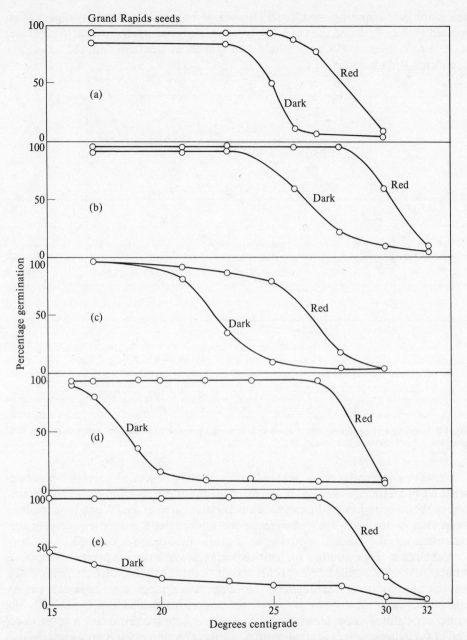

Fig. 6.2 Interaction between temperature and light in the germination behaviour of several different batches (a–e) of lettuce *var*. Grand Rapids seed. Note the same general pattern in all cases, but the great variability in detail (unpublished data of M. Al-Baghdadi and H. Smith).

under osmotic stress, whereupon dark germination is reduced more than light germination. Similarly, the existence of the photomechanism at supra-optimal temperatures can be demonstrated by applications of germination-stimulatory substances such as gibberellic acid. Figure 6.3 shows the effect of various concentrations of gibberellic acid on the *var*. Attractie lettuce seeds at 25°C, at which temperature the seeds are light-requiring, and at 30°C when germination is almost completely inhibited both in light and darkness.[124] Note that although gibberellic acid promotes dark germination at 25°C, both light and gibberellic acid are required to achieve significant stimulation at 30°C.

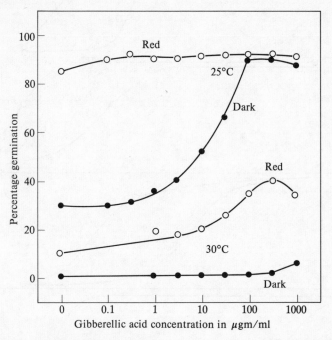

Fig. 6.3 **Interaction of gibberellic acid and red light in the germination of lettuce var. Attractie seed at 25°C and 30°C (after Frankland and Smith).**[124]

This would

These considerations seem to imply that both light-sensitive and temperature-sensitive processes are involved in the mechanisms of germination, and that whether one or the other process is dominant in a particular batch of seed depends on intrinsic properties of the seed itself. As mentioned above, these properties are probably determined by the conditions during maturation and storage of the seed.

Although a large number of light-requiring seeds have temperature responses similar to those described above, other response patterns are known. In many cases, photosensitivity occurs throughout the temperature range e.g., *Lepidium virginicum*.[438] Even in these cases, however, the underlying mechanism is probably the same, since although *Lepidium virginicum* requires light for germination at the lower end of the temperature range (in contrast to lettuce) germination is prevented by high temperatures in a manner exactly analogous to the behaviour of lettuce. One example, however, will suffice to show the variation possible in this type of light/temperature interaction. Figure 6.4 shows the response of *Amaranthus retroflexus* seeds to brief and continuous irradiation with white light, as affected by temperature.[472] It is clear that germination is stimulated by brief light treatment at all temperatures tested and also by continuous light between 25°C and 35°C. At low temperatures, however, seed germination is strongly inhibited by continuous illumination. This behaviour cannot at present be explained in a simple and convincing way.

No light requirement when chilled

6.6 Action spectra of the light effects

Early studies on the wavelength dependence of the photostimulation and photo-inhibition of seed germination led to the important generalization that, in both cases, red light (i.e., 600–680 nm) promoted germination, while far-red light (700–760 nm) inhibited germination. Thus, Flint and McAllister[120, 121] showed that red light stimulated the germination of photosensitive lettuce seeds while far-red light was inhibitory. Many other workers since then have confirmed this finding, both for lettuce and for other light-stimulated seeds. Similarly, Jones and Bailey[215] in 1956 showed that the germination of the white light-inhibited seed of the henbit, *Lamium amplexicaule* was also inhibited by far-red light and promoted by red light. Once again, these results have been confirmed by many other workers.

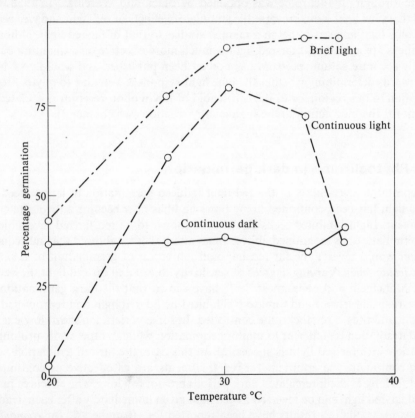

Fig. 6.4 Complex effects of brief and continuous white light treatments on the germination at different temperatures of *Amaranthus retroflexus* seed (after Khadman-Zahavi).[218]

The classic paper of Borthwick *et al.*[37] in 1952, as described in chapter 3, established that the effects of red light could be reversed by immediate subsequent irradiation with far-red light, and *vice versa*, demonstrating for the first time the existence of the red/far-red photoreversible pigment which we now call phytochrome. It must be considered proven, therefore, that phytochrome is the photoreceptor for both the light-stimulation and the light-inhibition of seed germination. The term photo-blastism has been used for many years now to cover the responses of seeds to white light. In view of the proven involvement of phytochrome, in both positive and negative photoblastism, it is reasonable to assume that the behaviour of seed depends solely upon its relative spectral sensitivity to the red and far-red regions. On this basis, the usefulness of the terms positive and negative photoblastism has recently been called into question.[411]

The only other spectral region known to have an effect on germination is the blue region (400–500 nm). Blue light has been reported to inhibit germination in the normally light-stimulated lettuce *var.* Grand Rapids and in the normally light-inhibited *Nemophila insignis*.[27] Other workers have stated that blue light can be either stimulatory or inhibitory to the germination of lettuce seeds, depending on the duration of imbibition prior to light treatment.[37, 113] Evidence that blue light may operate through phytochrome was obtained by Black and Wareing[27] in 1958 who found that blue light would reverse the stimulatory effects of red light and *vice versa*. Thus, blue light appears to act in a manner similar to that of far-red light, although all authors are agreed that far-red light is much more effective on a quantum basis. A really accurate action spectrum has not yet been published and would go a long way towards determining whether the blue light responses were due to phytochrome absorption in this region, or to the carotenoid or flavin photoreceptor postulated to account for the blue action in the high-energy reaction (see chapter 3).

6.7 Phytochrome in dark germination

The repeatable reversibility of the red light induced germination of lettuce seed by far-red light has been confirmed many times and has now become a favourite class experiment. Light-inhibited seeds can also be shown to be red/far-red reversible in their germination; as mentioned above, Jones and Bailey[215] showed that subsequent red light would reverse the far-red imposed inhibition of germination of *Lamium amplexicaule* seeds. Varying degrees of sensitivity to red and far-red light, however, occur. Mancinelli and colleagues[275, 276] have shown that the dark germination of some varieties of lettuce and tomato is inhibited by far-red light and repromoted by red light, and thus is phytochrome controlled. In some varieties, a short, low-energy far-red irradiation is sufficient to inhibit germination while in others only prolonged irradiation with far-red light is successful. In this case, the far-red irradiation need not be continuous; intermittent far-red treatments are as effective as continuous far-red, as long as the intercalated dark periods are not too long. The effects of intermittent far-red light can be reversed by red light given immediately after each irradiation treatment. Similar results have been reported for cucumber,[468] for *Nemophila insignis*,[358] and for *Amaranthus caudatus*.[230]

These observations are of fundamental importance to our present understanding of the role of phytochrome in germination and merit further analysis here. The fact that dark germination in these seeds can be inhibited by far-red light and repromoted by subsequent red light suggests immediately that Pfr is present in the dark-imbibed seeds and is responsible, or at least necessary, for germination. Removal of Pfr by far-red light thus prevents germination. On this view, the only difference between dark-germinating seeds and light-requiring seeds would be in the level of Pfr maintained in the dark seeds. Thus light-requiring seeds would have very low levels of Pfr – or none at all – and would require light to photoconvert Pr to Pfr. The dark-germinating seeds, on the other hand, would have considerable amounts of Pfr and could only be inhibited by light that converts Pfr to Pr. *See sheet (Kendrick)*

Is it possible, however, to explain on this basis the requirement of some varieties of seeds for continuous or intermittent far-red light? In order to explain such results it is necessary to postulate that Pfr is generated in darkness, following photoconversion of all pre-existing Pfr to Pr by far-red light. This hypothesis was originally put forward by Mancinelli and Borthwick[274] in 1964 in the following form:

1. Pfr is necessary for germination.
2. Seeds imbibing in darkness contain Pfr.
3. After removal of Pfr by far-red light, Pfr is formed again in darkness.

Thus in order fully to prevent germination it is necessary to continuously, or regularly, irradiate with far-red light to remove the Pfr being continuously formed.

6.8 The demonstration of Pfr in dark-imbibing seeds

The intriguing question immediately posed, of course, concerned the nature of the process whereby Pfr is generated in the dark. The favourite contenders were at first the rehydration of inactive Pfr during imbibition,[434] or the *de novo* synthesis of phytochrome in the Pfr form.[230] The first possibility seemed unlikely since sensitivity to far-red light slowly reappeared in darkness, after a brief irradiation with far-red light given many hours after the start of imbibition. This would mean that rehydration of phytochrome was occurring over a very long time period, in contrast with other seed proteins which tend to rehydrate rapidly during the initial phase of water uptake. The second alternative of continued *de novo* synthesis was a possibility, although the only direct spectrophotometric evidence on apparent phytochrome synthesis in etiolated plants indicated that phytochrome was produced solely in the Pr form.[72]

The obvious answer to these questions was to measure directly, by *in vivo* spectrophotometry, the levels of Pr, Pfr and total phytochrome in dry and imbibing seeds. Using the sophisticated instrument described on page 60 (Fig. 4.4) Spruit and various colleagues were able to demonstrate unequivocally the existence of Pfr in several dark-imbibing seeds, e.g., lettuce *var.* May Queen,[30] *Nemophila insignis,*[30] *Sinapis alba,*[30] *Amaranthus caudatus var. viridis,*[235] and *Cucumis sativus.*[408]

In lettuce, *Nemophila, Sinapis* and *Amaranthus* it was not possible to observe photoreversible absorbancy changes which could be attributed to phytochrome in the dry seeds. In *Cucumis,* however, phytochrome could be detected in the dry seeds, and in all species the levels of detectable phytochrome rose rapidly during the early hours of imbibition (Fig. 6.5).

All three groups of workers reported that a considerable fraction of the first-formed phytochrome in the imbibed seeds was in the Pfr form (i.e., lettuce, 40 per cent; *Amaranthus,* 25 per cent; *Cucumis,* 75 per cent). Since all the seeds investigated are dark germinators whose germination can be inhibited by far-red light, these direct observations of phytochrome fit in exactly with the predictions made from the physiological experiments.

The fact that phytochrome in dark-imbibing seeds can be largely in the Pfr form is in direct contrast with the known facts about phytochrome in etiolated seedlings where it is, as far as it is possible to measure, completely in the Pr form. Furthermore, in all the seeds investigated, there is evidence that a second period of phytochrome

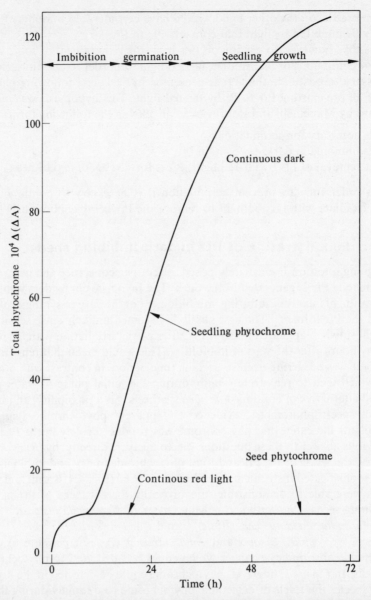

Fig. 6.5 The appearance of 'seed-phytochrome' during early imbibition, and 'seedling-phytochrome' during late imbibition, germination and seedling development in *Amaranthus caudatus*. Note that continuous red light prevents the appearance of seedling-phytochrome (probably through continuous Pfr destruction) but has no effect on seed-phytochrome (after Kendrick *et al*.).[235]

formation occurs much later in imbibition, and that this phytochrome is wholly in the Pr form. Figure 6.5 shows the extensive data of Kendrick *et al*. for *Amaranthus caudatus* in which an initial, relatively small amount of phytochrome is formed, followed by a massive rise after *ca*. 12 hours of dark imbibition.[235]

The second burst of phytochrome production can apparently be prevented by continuous irradiation with red or blue light. Continuous far-red light, on the other hand, has relatively little effect on the second burst of phytochrome formation. It is significant that although continuous blue and continuous far-red irradiation both, almost completely, prevent the germination of *Amaranthus caudatus* seeds, the level of phytochrome in the far-red light-treated seeds at germination time is approximately nine times as high as that in the blue light-treated seeds. Kendrick *et al.*[235] conclude from these observations that the 'seed-phytochrome' is different from the 'seedling-phytochrome' in that the former is stable in the Pfr form, while the latter is destroyed by the destruction process when in the Pfr form (see chapter 5). The stability of the 'seed-phytochrome' to red light, under which it should be about 80 per cent Pfr, is evident from the results in Fig. 6.5.

These results also provide possible support for the interpretation of many of the physiological responses of etiolated seedlings, on the basis of there being two pools of phytochrome – a small, 'active' pool, and a very much larger, inactive 'bulk' pool, as postulated by Hillman.[193] It seems quite reasonable to propose, as do Kendrick *et al.*[235] that their seed-phytochrome represents the 'active' pool of the etiolated seedling, while their seedling-phytochrome represents the 'bulk' pool (see also page 106).

6.9   Reappearance of Pfr in darkness

These results, then, provide convincing proof of the hypothesis of Mancinelli and Borthwick[274] that Pfr exists in dark-imbibed seeds. They do not, on the other hand, explain the requirement for intermittent far-red light for the complete inhibition of germination of several such dark-germinating seeds. This requirement can probably be best accounted for on the basis of the formation of further Pfr in darkness following a short irradiation with far-red light, which should photoconvert all pre-existing Pfr to Pr.

One possible way to account for this would be to say that *de novo* synthesis of phytochrome in the Pfr form continues throughout the whole period of imbibition. The results in Fig. 6.5 do show that continued phytochrome production occurs throughout imbibition. Only the first formed phytochrome – the seed-phytochrome – however, has a significant proportion in the Pfr form. In all cases, the seedling-phytochrome produced in the later stages of imbibition is wholly in the Pr form.[28, 30, 235, 408] It is therefore not easy to explain the requirement for intermittent far-red light for complete suppression of germination, on the basis of continual *de novo* synthesis of Pfr.

A most intriguing possibility, on the other hand, was raised by Boisard *et al.*[30] who discovered that in imbibed seeds of lettuce *var.* May Queen, and *Nemophila insignis*, Pr produced from Pfr by far-red light was capable of rapid conversion, in darkness, back to Pfr. This process, which they named 'inverse dark reversion' (see also page 99) since it is exactly the opposite of the long accepted dark reversion of Pfr to Pr in etiolated plants, has since been confirmed for *Amaranthus caudatus*,[235] and for *Cucumis sativus*.[408] Figure 6.6 shows the time-course of inverse reversion in *Amaranthus* seeds.[235]

The capacity for dark reversion of Pr to Pfr is a property only of the seed-phytochrome and is not found in seedling-phytochrome. Even so, seed-phytochrome appears to retain the capacity for inverse reversion throughout imbibition and at least up to germination time. These results, then, provide the possibility that inverse reversion of Pr to Pfr in darkness may be the cause of the requirement for continuous or intermittent far-red light for the complete suppression of germination.

6.10 Interaction with hormones and the rapid action of phytochrome in seed germination

Impotent!

The initiation of germination in light-sensitive seeds normally takes place several hours, or even days, after light treatment. Furthermore, in light-sensitive varieties of lettuce, it has been shown that far-red reversibility of the red light induced stimulation of germination can be possible very many hours after the red light treatment.[39] This has generally been taken to indicate that the presence of functional Pfr for long periods is necessary for germination. Experiments on the interaction between phytochrome and certain hormones, however, seem to indicate that the situation is not quite as straightforward as it seems.

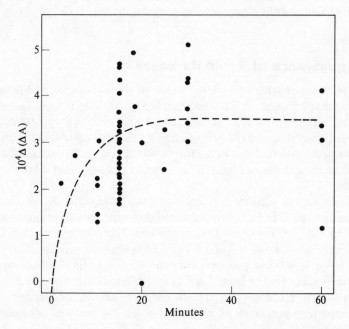

Fig. 6.6 The time-course of inverse dark reversion of Pr and Pfr in *Amaranthus caudatus* (after Kendrick et al.).[235]

Cytokinins were the first hormonal substances found to stimulate the germination of light-sensitive lettuce seeds.[285,286] It is now known that, although small increases in germination percentage can be induced by treatment of dark-imbibed lettuce seeds with kinetin, the most striking effects are observed when sub-threshold amounts of both red light and kinetin are given together.[258] Applications of gibberellins, on the other hand, will cause germination of lettuce *var*. Grand Rapids seed in darkness, but again, synergistic interactions between sub-threshold amounts of red light and gibberellic acid occur.[203,222] Thus in both cases, maximum effects are obtained when both red light and hormones are given.

This is an important point since it could be suggested that phytochrome may operate by switching on the production of gibberellins or cytokinins (or both), which then go on to initiate the hydrolysis of stored materials, and the cell expansion processes, which constitute germination. The synergism between light and the hormones, however, indicates that phytochrome and the endogenous hormones operate along separate paths, both of which are required for ultimate germination.

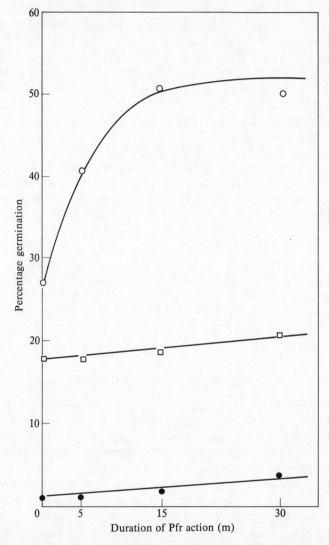

Fig. 6.7 Interaction between Pfr and sub-threshold concentration of gibberellic acid. Lower curve, water only; upper curve, 5 μg/ml GA₃; middle curve, calculated curve for additive effect of Pfr and 5 μg/ml GA₃ (from Bewley *et al.*).[18]

Utilizing the synergism between Pfr production and exogenous gibberellin in lettuce *var*. Grand Rapids seeds, Bewley *et al.*[17,18] were able to carry out some ingenious experiments which demonstrated a very rapid action of phytochrome in seed germination. Since Pfr can be reconverted to Pr by far-red light, it is possible to control accurately the period of time during which Pfr is available in the seed. Allowing Pfr to be present in water-imbibed seeds for short periods (5–30 minutes) had no significant effect on subsequent germination. If very low (sub-threshold) concentrations of gibberellic acid were present however, even 5 minutes of Pfr action caused a very marked increase in ultimate germination percentage (Fig. 6.7). Similar results were observed with six other gibberellins, indicating that Pfr production does not cause the synthesis of those compounds. The fact that phytochrome action occurs very rapidly after the photoconversion has been confirmed in other phenomena, and is of great importance to our general understanding of the way phytochrome functions, as has been discussed in chapter 5. As far as germination is concerned, these results appear to indicate that phytochrome is only one of a variety of parallel and mutually interacting mechanisms which together are responsible for the switching on of germination. The other mechanisms probably include the release, or activation, of gibberellins and cytokinins and processes which enable the bypassing of the phytochrome steps by low temperature.

Since germination probably involves the simultaneous action of several independent, but interacting, stepwise sequences, it may ultimately be necessary to use the methods of the systems analyst to achieve any degree of satisfactory understanding. A plausible approach, for example, may be to use a variant of critical path analysis, which seeks to identify the separate indispensable processes involved in the manufacture of a new product, and to show how certain steps can occur only upon the completion of other steps. In seed germination under normal conditions, we can identify the photoconversion of seed-phytochrome, the production or release of gibberellins and cytokinins, and the maintenance of an optimum temperature, as three of the many critical paths towards germination. We must also assume, however, that under other conditions, e.g., massive exogenous applications of gibberellins, or low temperature, mechanisms exist for overriding other critical paths, e.g., the phytochrome path.

6.11 Wider implications of the seed phytochrome concept

The discovery of seed-phytochrome and its distinctive properties leads us inevitably to re-examine some of the widely-accepted concepts – even dogmas – of phytochrome action. For example, it has long been universally accepted that Pr is stable and not subject to any conversions other than photochemical ones. Also, that Pfr is unstable and rapidly degraded by the process of destruction. The properties of seed-phytochrome are almost the exact reverse of these, but it must be remembered that the dogmas referred to above were arrived at by using relatively insensitive instrumentation in investigations of bulk phytochrome in etiolated seedlings and extracts.

At the present time, it seems likely that seed-phytochrome may be representative of the active fraction of the total phytochrome of all plants, and thus it is necessary to consider whether or not the existing hypotheses of phytochrome action are consistent with our knowledge of seed-phytochrome properties. The evidence from the germination and spectrophotometric experiments described above has been taken to indicate that Pfr is necessary, and responsible, for germination. Thus, repeated removal of Pfr prevents germination by the removal of an essential, active component of the chain leading to germination. This concept can be summarized in the following way; Pfr is thought to catalyse in some way the formation of a product Z, which is essential for germination, from a substrate S:

$$Pr \underset{\text{far-red}}{\overset{\overset{\overset{\text{dark}}{\text{red}}}{\rightarrow}}{\rightleftharpoons}} Pfr + S \rightleftharpoons PfrS \longrightarrow Z$$
$$\downarrow$$
$$\downarrow$$
$$\text{Germination}$$

The rapid action of phytochrome is not consistent with this scheme, as it is written here. On the other hand, it is not necessary to assume that Pfr acts in an enzymatic fashion as depicted. As described in chapter 5, compelling evidence exists that phytochrome somehow acts through changing, either selectively or generally, membrane permeability. If it is assumed that Pfr acts on important membranes in this way, then the immediate interaction of gibberellin with Pfr can be explained on the basis of immediate gibberellin interaction with substances released through the membrane. Thus, in a modified form, the scheme above is certainly tenable.

To show that a particular model of phytochrome action is consistent with a set of data does not constitute proof of that model. At the present time, all models of phytochrome action are hypothetical, and it seems likely that a great deal of research will be necessary before we will have sufficient positive evidence to dignify any of these hypotheses with the title 'theory'. It is necessary, as always, for the reader to keep an open mind and to continually question the ideas placed before him.

In this speculative, but critical, frame of mind we may with advantage consider some of the general problems of the role of light in seed germination in an attempt to determine whether or not they can be accounted for on the basis of phytochrome. For example, is it possible now to postulate a mechanism for the assumption of skotodormancy and photodormancy in seeds kept in unfavourable light conditions? Since both forms of induced dormancy occur when imbibed seeds are kept under conditions unfavourable for the formation of Pfr, it would seem likely that here again phytochrome is the key.

Two alternative hypotheses offer themselves as candidates. It is possible that the substance with which Pfr combines (i.e., S in the scheme above or X^I in Fig. 5.20) is unstable, and that its continued presence depends upon continued synthesis. With time, the capacity for its synthesis may gradually decline as, perhaps, substrates are used up, and ultimately the level of S may become too low for interaction with Pfr. Thus, after a considerable period of time, even though Pfr may then be produced by the correct light treatment, it will not be able to combine with S, and thus germination will not occur.

The alternative idea is that the Pr of seed-phytochrome is itself unstable, and continued maintenance of seed-phytochrome in the Pr form leads to its ultimate loss. Thus, subsequent irradiation with red light will not cause the formation of sufficient Pfr to switch on the germination processes. At first sight, it would seem possible to assess the credibility of this second alternative by *in vivo* spectrophotometry, but it must be remembered that during the dormancy-induction period massive quantities of seedling-phytochrome will have appeared, thus masking any absorbance changes due to seed phytochrome. On the other hand, treatments which can sometimes remove skotodormancy and photodormancy (e.g., gibberellins, low temperature) also lead to a resumption of red/far-red photoreversibility, in many cases.[110]

A final problem worthy of consideration is the question of why certain batches of a particular seed should be light-requiring while others germinate fully in darkness. It seems most likely, as implied earlier in this chapter, that environmental conditions, particularly light quality during seed maturation, are very important in deciding the subsequent germination behaviour. McCullough and Shropshire[270] in 1970 working with *Arabidopsis thaliana* showed that dark germination percentages for seeds matured under cool white fluorescent lamps averaged 45 per cent, while seeds matured under incandescent illumination gave no germination in the dark. Cool white fluorescent sources omit very little radiation in the far-red region (i.e., above 700 nm) while a large proportion of the visible irradiation emitted by incandescent lamps is in the far-red region. Thus, the seeds which matured under light qualities maintaining phytochrome predominantly in the Pr form had an absolute requirement for red light for germination, while seed matured with phytochrome partly in the Pfr form showed partial germination in the dark.

Although actual measurement of phytochrome in the seeds was not possible, the correlations are very striking, and it seems most likely that the state of phytochrome at the time of seed drying determines the future germination response. Modifications do occur during storage and after-ripening of the seed, but the principal factor affecting light sensitivity appears to be the conditions of maturation. These relationships may well be of important significance for the ecology of seed germination.

7

The photocontrol of
seedling development

7.1   Etiolation and de-etiolation

When seedlings of higher plants are grown in complete darkness they become
etiolated. The term etiolation is derived from the French *étioler*, which means *to
grow weak*, *to grow pale*, and is generally descriptive of the most obvious common
symptoms of etiolation, i.e., the lack of production of chlorophyll and the excessive
spindly growth of the stems. The term etiolation can only be used in a general sense,
however, since the specific morphogenic effects of the deprivation of light differ
between the various classes of higher plants.

The first recorded comments on the phenomena of etiolation were those of the
great seventeenth-century English botanist, John Ray,[350] and experimental observations on etiolation were published as early as 1754 by Bonnet.[32] Much experimental
work was carried out in the latter half of the nineteenth century, and in 1903 Mac-
Dougall[272] wrote a monograph on the anatomical and morphological effects of the
deprivation of light, which remains to this date a most comprehensive treatise well
worthy of study. MacDougall precisely documented the great variation in etiolation
effects between different plants, a point which is often overlooked today.

The most common morphogenic effects of growth in complete darkness are the
extreme elongation growth of the internodes and the suppression of leaf development.
This pattern is only really valid, however, for some dicotyledonous plants, since in
monocotyledonous seedlings leaf growth is commonly not inhibited by darkness,
and in some cases, according to MacDougall, is even stimulated by darkness (e.g.,
Narcissus). In many, but not all dicotyledonous seedlings, etiolation results in the
formation of pronounced plumular, or hypocotylar hooks (see the Frontispiece).

The changes wrought by etiolation are not restricted to these gross morphological
effects. MacDougall considered one of his major conclusions to be the finding that
tissue differentiation was much retarded during etiolation, particularly with regard
to xylem differentiation. This aspect of etiolation has also been somewhat neglected
in recent years, although in the twenties Priestley and coworkers carried out intensive
investigations of tissue differentiation in *Vicia* and *Pisum* stems, as affected by growth
in darkness or light.[340-342] They showed that during etiolation, an endodermis layer
with a typical Casparian strip was often laid down in the stem tissues in place of the
starch sheath formed during normal growth in the light. Priestley[341] further postulated that the development of an endodermis results in the limitation of growth
activity to those regions contained within the strip, accounting for the inhibition of

leaf development. He believed that the lack of light caused the conversion of carbohydrates to lipids, which were deposited as the Casparian strips, and that increased lipid content in the walls of the cells immediately below the meristem resulted in decreased translocation of nutrients to that meristem, thus inhibiting organogenesis.

These findings were later found not to be, in any way, general for other etiolated seedlings, and it was even suggested that *Vicia* and *Pisum* are atypical in this respect.[31] If this is true, it is perhaps unfortunate that both *Vicia* and *Pisum* have become such popular experimental organisms among photomorphogenecists.

In later work,[146, 240, 409] MacDougall's assertion, that the lack of light inhibited xylem development, has been substantially confirmed. Kleiber and Mohr[240] moreover, have shown that light acts in a quantitative way to accelerate the differentiation of tracheids and vessel elements, but has no qualitative effect on the overall pattern of differentiation.

Etiolated seedlings have been a favourite experimental tool for the investigation of photomorphogenesis, since transferring such seedlings to daylight causes a transition to the developmental pattern of fully light-grown plants. Thus the young, immature tissues undergo profound developmental changes which can be finely controlled by manipulation of the light treatments, and can therefore be analysed with great precision. Probably the most exhaustive analysis of the range of photoresponses exhibited by the etiolated seedlings of a single species is that carried out by Mohr and his colleagues at Freiburg, on *Sinapis alba*, the white mustard. A list of the known photomorphogenic responses of *Sinapis* is given in Table 7.1. It is clear that treatment of an etiolated *Sinapis* seedling with light initiates an enormously wide range of simultaneously occurring processes which together constitute a complex,

Table 7.1 Developmental and biochemical processes brought about by light treatment of etiolated *Sinapis alba* seedlings (from Mohr[297]).

Inhibition of hypocotyl lengthening
Inhibition of translocation from the cotyledons
Enlargement of the cotyledons
Unfolding of the lamina of the cotyledons
Hair formation along the hypocotyl
Opening of the hypocotylar hook
Formation of leaf primordia
Differentiation of primary leaves
Increase in negative geotropic reactivity of the hypocotyl
Formation of tracheid elements
Differentiation of stomata in the epidermis of the cotyledons
Formation of plastids (from etioplasts) in the mesophyll of the cotyledons
Changes in the rate of cell respiration
Synthesis of anthocyanin
Increase in the rate of carotenoid synthesis
Increase in the rate of ascorbic acid synthesis
Increase in the rate of protochlorophyll synthesis
Increase of RNA synthesis in the cotyledons
Decrease of RNA contents in the hypocotyl
Increase of protein synthesis in the cotyledons
Changes in the rate of degradation of storage fat
Changes in the rate of degradation of storage protein

overall, developmental change. This overall change is called *de-etiolation*, a rather unfortunate term for which there is no effective synonym.

De-etiolation is a complex phenomenon not only in the range and extent of the specific developmental changes involved, but also in the varieties of photoreactive mechanisms which are implicated. As will be seen, complete de-etiolation requires light perception via phytochrome and the high-energy reaction, perception via specific photochemical reactions in chlorophyll synthesis, and also probably perception via photosynthesis. It is not possible at the present time to present a fully integrated picture of the roles of light in seedling development; there are still many areas in which the level of knowledge and understanding is very meagre. In consequence, this chapter concentrates on those partial processes of de-etiolation which can be outlined fairly confidently, at least in principle. In the subsequent chapter, biochemical changes accompanying de-etiolation are discussed.

7.2 The pattern of light-mediated developmental changes

It is a matter of common observation that internode growth in both dicotyledons and monocotyledons is inhibited by light, whereas leaf expansion (at least in dicotyledons) is stimulated by light. Went[456] in 1941 observed that light treatment of pea seedlings inhibited the growth of one internode while stimulating that of the internode above. He concluded that this was due to a compensatory growth mechanism, in which the younger internode was able to grow more quickly because the light-inhibited lower internode no longer monopolized certain substances required for stem growth, which were produced in the cotyledons or roots. This concept of compensatory growth describes, but in a rather obscure way, what has become an important generalization – young stem tissues are, in fact, stimulated in their growth by light, while the older tissues only are inhibited.

This fact was first pointed out by Thomson[428] in 1950, when she showed that the effects of light on the growth of the various organs of the etiolated oat seedling depended on the stage of development of these organs. Thus, the growth of very young coleoptiles, internodes, or leaves was in all cases increased by light while the growth of almost mature coleoptiles, and internodes (but not leaves) was inhibited by light. Thomson concluded that all organs go through a specific sequence of developmental stages, consisting of a stage of cell division, a stage of cell enlargement, and finally a stage of cell maturation. On this hypothesis, it is possible to rationalize the differential effects of light on organs of different ages by assuming that the prime effect of light is to *accelerate* all phases of the division-enlargement-maturation sequence.

Thus in older tissues, overall growth inhibition can be considered as an accelerated transition from the cell enlargement to the cell maturation phase, while in younger tissues, stimulation is due to an accelerated transition from the division to the enlargement phase. This behaviour is shown dramatically in Fig. 7.1, where the growth curves of oat coleoptiles are shown in continuous darkness and continuous light. It is clearly obvious that light accelerates the time of onset of extension growth, yet similarly hastens the cessation of extension growth, due to the accelerated maturation and differentiation of the cells.

Thomson[429] considers it likely that leaves behave in an exactly analogous manner, except that they never, when grown in darkness, pass beyond a stage corresponding to the earlier part of internode development, and thus light never elicits inhibition responses.

The significance of this generalization is great. If light is considered to have a solely inhibitory effect on stem growth and an opposite, stimulatory effect on leaf growth, we are forced to search for two different mechanisms of light action. This view is prevalent in much of the literature, and in fact a formal categorization of stimulatory and inhibitory responses as 'positive' and 'negative' photoresponses has been proposed by Mohr.[295] Positive photoresponses are thought to be due to

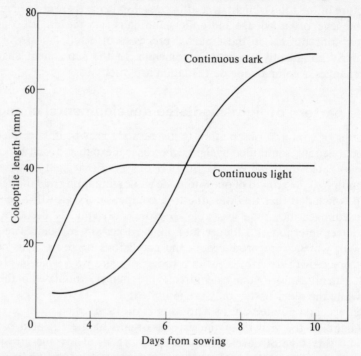

Fig. 7.1 The dual effect of light on coleoptile growth. Light stimulates growth in the early stages of coleoptile development, but inhibits later. Similar curves can be constructed for stem growth but not for leaf growth (after Thomson).[429]

the light-mediated activation of key genes, while negative photoresponses are considered to be due to the repression of other key genes. The evidence for these proposals rests on the fact that actinomycin D (an inhibitor of DNA-dependent RNA synthesis) and puromycin (an inhibitor of protein synthesis) prevent the 'positive' photoresponses of hypocotyl hook opening, cotyledon expansion, lamina unfolding, and anthocyanin synthesis in etiolated *Sinapis* seedlings, whereas these substances do not reverse the 'negative' photoresponse of inhibition of internode extension.[296] It is also true, however, that actinomycin D has no inhibitory effect on the light-stimulated increase in ascorbic acid level in *Sinapis* seedlings, a response which has all the characteristics of the postulated 'positive' photoresponses,[380] and thus the general usefulness of the concept of 'positive' and 'negative' photoresponses seems in considerable doubt.

Development within the apex appears to be only slightly affected by the lack of light, and even then the effects are purely quantitative. Thus in *Pisum* and *Vicia*, light causes a small acceleration in the rate of inception of new leaf primordia, although no changes in the pattern of primordium production were observed.[61, 431] In peas, the first detectable effect of light is to be seen in the internode separated from the apex by two leaf primordia; here the stem develops thicker in the light than in darkness, and cortex differentiation starts earlier. Below the fourth leaf from the apex, internode elongation is stimulated by light. Light reduces final cell size and number in the internodes, although it does not alter the pattern of primary differentiation. Thomson and Miller[432] concluded that the earlier non-polar phase of primordium production, cell differentiation and isodiametric cell growth is stimulated by light, whereas the later, strongly polarized phase of internode elongation, is inhibited by light. Thomson and Miller's intensive investigations of the behaviour of the pea plant lead to the conclusion that the great morphological differences between etiolated and light-grown seedlings are due to quantitative effects on the rate and extent of the developmental processes, and not to qualitative changes in the overall pattern of development.

7.3   Analysis of light requirements

In 1949, Parker et al.,[324] at Beltsville, showed that the action spectrum for the light-stimulated increase of leaf area in etiolated *Pisum* seedlings had a peak in the red region similar to the action spectrum for photoperiodic flower induction. Borthwick et al.[35] in 1951 found a similar action spectrum for the inhibition of internode elongation in both normal barley and albino barley seedlings. These results suggested that phytochrome was the photoreceptor for both the stimulation of leaf expansion and the inhibition of internode extension, and this conclusion was proved correct when Downs[89] showed that the photocontrol of the two processes in kidney bean seedlings was red/far-red reversible. This conclusion has been confirmed since then by very many investigators and, as reported in chapter 3, studies of the photocontrol of seedling development have proved invaluable to our gradual understanding of the functions of phytochrome.

Photoreactive response systems other than phytochrome, however, are undoubtedly involved in de-etiolation. In 1957, Mohr[292] found that extension growth in *Sinapis* seedlings could be inhibited by the high-energy system – i.e., irradiation with far-red or blue light of relatively high energy, and for relatively long periods of time, would cause inhibition of hypocotyl growth. Furthermore, the maximum inhibition that could be achieved with this sytem was greater than the maximum achieved by the single photoactivation of phytochrome by red light. As is described in chapter 3, the far-red band of action in the high-energy reaction is now thought to result from the simultaneous excitation of Pr and Pfr; the photoreceptor for the blue band of action, on the other hand, has not yet been identified.

Some of the evidence that the effects of high-energy blue and far-red light on seedling growth are exerted through different mechanisms is also described in chapter 3. This evidence is supported by detailed investigation of the time courses of the two photoreactions. In gherkin hypocotyls, for example, both blue and far-red light inhibit extension growth, but detailed kinetic experiments show major differences in the time courses of the effects.

Using a sophisticated automatically recording auxanometer, Meijer[283] was able to record continually the growth rate of the gherkin seedlings with a high degree of precision. Transfer from continuous darkness to continuous red, far-red, and blue light of closely similar irradiances, in each case, resulted in a decrease in growth rate (Fig. 7.2). With red and far-red light, lag phases of 30 minutes and 5 minutes. respectively, preceded the drop in growth rate, which was even then relatively gradual. With blue light, on the other hand, an instantaneous cessation of growth was recorded. These results indicate very strongly that the effect of blue light is mediated through a different photoresponsive reaction mechanism from that mediating the red and far-red light effects.

Further evidence for this view comes from the observations of Evans et al.[108] that lettuce seedlings grown in complete darkness for more than 60 hours lose their sensitivity to far-red, while retaining their sensitivity to blue light. Such conclusions can be derived erroneously, however, as was shown by Häcker et al.[160] also working

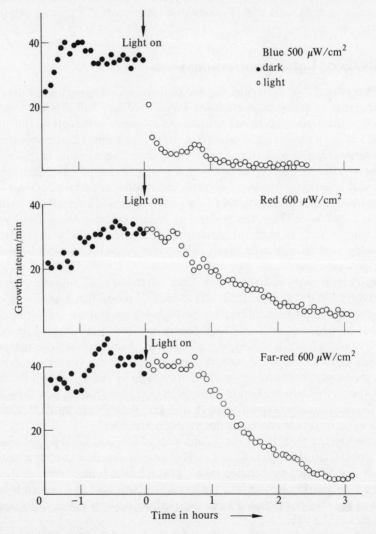

Fig. 7.2 The rapid inhibition of stem growth by blue light in comparison with the relatively slow effects of red and far-red light (from Meijer).[283]

144   Phytochrome and photomorphogenesis

with lettuce hypocotyls. The overall growth of hypocotyls of seedlings between 60 and 84 hours after germination was the same in red light and darkness, implying that red light had no effect. On the contrary, however, it was demonstrated by cytological observations that red light slightly inhibits cellular lengthening in the middle part of the hypocotyl, while slightly stimulating extension in the upper part, both effects exactly cancelling each other out in a manner consistent with the Thomson hypothesis. This type of behaviour would be undetectable by measurements of overall growth, leading to the interpretation that photosensitivity had been lost.

It is thus clear that at least three separable photoreactions are involved in the complex process of de-etiolation – i.e., the low-energy phytochrome system, the high-energy phytochrome system, and the high-energy blue system. It is also known, as is described in chapter 8, that yet another photochemical reaction is involved in the pathway of chlorophyll synthesis, and when it is realized that later growth of the illuminated seedling depends on the products of photosynthesis, then the difficulties of integrating the available information become evident.

7.4 Light effects on light-grown seedlings

Etiolated seedlings are plants which are growing under abnormal conditions, and therefore the possibility exists that the photoreactions comprising de-etiolation may be restricted to dark-grown plants. This is not the case, however, and it is now known that all three photoreactions described above play a role in the development of the 'normal' seedlings, i.e., the seedling growing under normal daylight conditions. The sensitivities of etiolated and light-grown plants in the different reactions are not always the same, but it is abundantly clear that the photoreactions are operative. These considerations are important for our understanding of the normal processes of plant development, and also for our understanding of the ecological role of photomorphogenesis.

The methods used in investigations of the roles of the different photoreactions in normal plant growth consist mainly of growing plants under photoperiods of standard white light (e.g., 12 or 16 hours of white light from a fluorescent source) and supplementing this with shorter periods of light of restricted wavelength regions. Other methods involve growing plants for long periods in restricted spectral regions. This sort of work has obvious commercial interest, and has been extensively pursued by various agricultural research laboratories, amongst the most productive being the Landbouwhogeschool (Agricultural University) at Wageningen, and the Philips Research Laboratories at Eindhoven, both in the Netherlands.

Although it is difficult to detect phytochrome in light-grown tissues, it is nevertheless easy to demonstrate photomorphogenic phenomena that are patently phytochrome-mediated. For example, Downs et al.[91] in 1957 found that stem elongation of a number of species could be controlled by brief supplemental treatments with red or far-red light immediately after the end of the daily photoperiod. The responses were red/far-red reversible, stem elongation being greater when the final irradiation was with far-red light. These effects are of high magnitude, since in some species exposure of light-grown plants to brief far-red light of low irradiance, at the end of a daily photoperiod, causes increases of 2 or three times in the growth rates of individual internodes.

Most investigators appear to consider these supplemental red/far-red light effects to be phytochrome-mediated, and to be similar in nature to the accelerated cellular maturation of the Thomson hypothesis. There are, however, some intriguing species differences. Nakata and Lockhart,[313] for example, working with *Phaseolus vulgaris*, found that supplemental far-red light stimulated both cell division and cell elongation equally, whereas Le Noir[259] working with *Phaseolus acutifolius* accounted for all the stimulation on the basis of increases in the rate and duration of cell elongation only. Obviously, with such complex experimental treatments, it is possible that differences of this sort may be due to slight inconsistencies in the methods used; for example, different energies of red and far-red light, different ages of plant material, etc. On the other hand, the diversity of photomorphogenic responses so well known to Mac-Dougall may be the cause of these differences. In any case, the existence of such different responses makes it difficult, and dangerous, to attempt the formulation of a unified pattern of the mechanisms involved.

When plants are grown for long periods under restricted spectral regions, the resulting growth pattern appears to be determined by three factors:[282]

1. whether the plants were previously grown in darkness or in white light;
2. the spectral region used; and
3. the irradiance used.

Table 7.2 shows the effects on growth of 16 hours per day of red, blue, or green light, with or without simultaneous far-red light, in five different species. The seed-lings used had all been growing under white fluorescent lamps prior to the treatment. It is clear from this table that marked inter-specific differences occur, although in all cases red and green light appear to have virtually equal effects on extension growth.

Table 7.2 The effects of different spectral regions on growth.
The influence of an irradiation (16 hours per day) with red (16R), green (16G), and blue light (16B), with or without far-red light (FR), on the elongation of light-grown plants. Irradiance of visible radiation was 475 μW/cm^2 and of far-red light 120 μW/cm^2. Length given in mm, N days after the beginning of the treatment (second column) (from Meijer[282]).

Plant species	N (days)	16R	16RFR	16G	16GFR	16B	16BFR
Gherkin (hypocotyl) 5 days*	8	23	36	26	101	79	81
Mirabilis jalapa (First internode) 14 days*	30	182	179	174	188	128	153
Salvia occidentalis (internode) 20 days*	30	42	60	43	82	64	90
Tomato (hypocotyl) 14 days*	11	27	29	24	54	28	42
Tomato (First internode) 14 days*	15	69	74	62	63	38	38

* Age of the plants at the beginning of the treatments.

A further generalization is that far-red light, when it has an effect at all, leads to an increase in extension growth. Comparisons of the effects of blue and red light on the growth of the different species, however, show clearly that in some cases, e.g., gherkin, and *Salvia occidentalis*, blue light leads to greater growth, while in other species, e.g., *Mirabilis jalapa*, and the internodes of tomato, blue light leads to less growth.

To complicate matters even further, the relative effectiveness of equal irradiances of red and blue light depends on whether the plant has been previously grown in white light or in darkness. In tomato seedlings, grown in white light and transferred at various times to red, green or blue light, those internodes which had already received white light became longer in blue light than in red or green light. Those internodes which developed later, and thus had never been exposed to the white light, were longer in red and green light, than in blue light.[282]

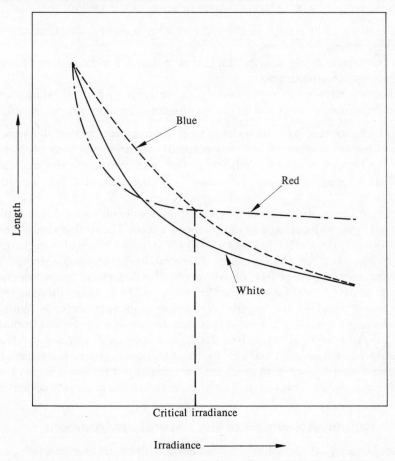

Fig. 7.3 Generalized response versus irradiance relationships for the effects of continuous blue, red and white light on seedling growth (after Meijer).[282]

These results fit in, of course, with there being at least two photoreceptors, one active in red (and possibly green) light, and one active in blue light, and differing in their relative quantum effectiveness in etiolated and light-grown material. This conclusion is also borne out by the observation that the response versus irradiance curves for red and blue light are of a different shape, and overlap[282] (Fig. 7.3). If it is assumed that the point of overlap (which is essentially the point at which equal irradiances of blue and red light give equal responses and is termed critical irradiance by Meijer) differs from species to species, then the marked variation in comparative effects shown in table 7.2 can be accounted for.

The involvement of these different photomorphogenic reaction systems in the growth of plants under 'normal' conditions obviously has important practical applications, especially for the growth of plants in greenhouses, and controlled environment facilities, with artificial illumination. Much investigation has gone into the design of lamps to give optimum growth of economically important glasshouse crops, such as tomatoes, and an excellent guide to the practical aspects of this subject is contained in the books by Canham[67] and by Bickford and Dunn.[19] In the design of a suitable lamp, due consideration must be given to the need to:

1. provide sufficient energy in the red and blue regions for optimum rates of photosynthesis;
2. give a spectral energy distribution that is optimum for the pattern of development that is required; and
3. to keep the infra-red energy as low as possible to cut down overheating problems and the consequent expense of air-conditioning for temperature control.

The white fluorescent lamp is almost ideal in these respects, in that all its energy is emitted within the visible region, thus markedly reducing the heat problem, yet maintaining high photosynthetic efficiency. This is in contrast to the incandescent lamp which, although cheap and convenient, is deficient in blue light and high in infra-red light.

Fluorescent lamps, for the most part, cut-off at about 700 nm, and thus emit very little far-red light, with consequent morphogenic effects. These effects are not always undesirable, since a lack of far-red light usually results in inhibition of stem elongation, leading to the short, stocky, seedlings favoured by nurserymen. Meijer,[284] for example, has compared the effects of two very similar fluorescent lamps, which differ in the relative amounts of far-red light they emit, and he has found the far-red rich lamp to produce undesirable excessive elongation of gherkin seedlings. With some plants, on the other hand, it is known that the addition of some far-red from incandescent lamps to the light emitted by fluorescent lamps will increase productivity when measured as dry-weight yield.[457] In recent years, several new fluorescent lamps have been produced which emit much higher energies in the far-red region leading, at least for some species, to considerable apparent increases in gross productivity.[426]

7.5 Photomorphogenesis in the natural environment

Plants are not adapted to continuous irradiation with monochromatic light, nor do they experience brief irradiations with red and far-red light. If we reject the unlikely hypothesis that red/far-red photoreversibility evolved in order to make the physiologist's task more easy, we are obliged to consider the adaptive value to plants of having the phytochrome system. It is clear that detection of light is of great importance to a germinating seed, or developing seedling, so that its development may be appropriate for growth below or above the soil surface. Thus, seeds whose germination is inhibited by light would not germinate unless they were buried beneath the soil. On the other hand, seeds whose germination is stimulated by light would remain dormant for lohg periods in the soil, and would germinate only when the soil was disturbed, and they were exposed to light. This behaviour would be expected to be of great survival value to the species since it would have the effect of spreading out the times of germination of the seed, thus providing a greater chance that at least some seed would germinate during favourable climatic conditions. Similar arguments can be extended concerning etiolation and de-etiolation. It is clearly advantageous to the young seedling to reach the surface of the soil as quickly as possible so that it may begin obtaining its energy from sunlight by photosynthesis. The peculiar developmental adaptations that characterize etiolation, i.e., extreme stem elongation, and reduced leaf extension, clearly confer these advantages.

It is obvious, then, that photosensitivity confers adaptive advantage to plants at certain stages of their life-history. These considerations, however, do not provide any clues as to the selection pressures which have led to the evolution of the highly complex phytochrome system as the effective agency of light sensitivity. This is, in fact, one of the most important unanswered – and largely unasked – questions concerning phytochrome. In order to approach an answer, it is necessary to compare the properties of phytochrome with the important parameters of the natural environment, and to determine whether any meaningful correlations exist.

Plants in the natural environment are exposed to long periods of irradiation with light from a broad-spectrum source – i.e., the sun. Consideration of chapters 3 and 4 will show that under these circumstances, any phytochrome present in the plants will be driven to a photostationary state (i.e., Pfr/Ptotal) determined by the proportions of red and far-red wavelengths present in the incident irradiation. For monochromatic irradiation, the characteristic photostationary states can be readily determined,[167] and from these data, together with the spectral composition of the radiation incident upon the plant, it is possible to calculate the expected Pfr/Ptotal for any radiation source. (The discovery[231, 344] that photoconversion intermediates build up under high irradiances somewhat complicates this situation, since if much of the total P is in an intermediate form the proportions of Pr and Pfr might be expected to differ.)

When daylight, which has a maximum energy per unit wavelength at *ca.* 420 nm (see Fig. 2.5 for spectral distribution), passes through a vegetation canopy, much of the red and blue wavelengths are absorbed by the chlorophyll in the leaves. The absorption spectrum of a typical leaf is shown in Fig. 7.4 and the spectral distribution of the light beneath the canopy ('canopy-light') is shown in Figs. 7.5 and 7.6 for mid-day, and for sunrise and sunset.

It can easily be seen that canopy-light is relatively very rich in far-red light compared to unfiltered daylight. The vegetation acts as a superb far-red filter (in fact it has been shown to reverse the red light potentiation of lettuce seed germination[25]) and thus we must expect the Pfr/Ptotal of plants growing within the canopy to be very different from those growing in open ground. Some calculations and observed results are shown in Table 7.3 proving this to be the case. Note also there is considerable fluctuation on a daily basis.

It is quite possible that this major change in Pfr/Ptotal within the shaded plant could bring about a changed pattern of development appropriate to the ecological situation. That this may well be so is indicated by the experiments referred to on page 145 in which extra far-red light was given to plants growing in otherwise white

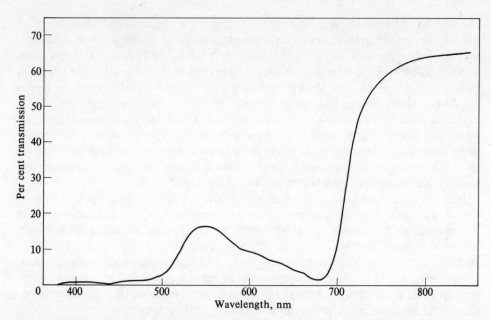

Fig. 7.4 Transmission spectrum of a young leaf of sugar-beet. (The transmission in the far-red region is higher than that of a more mature leaf) (from Smith).[402]

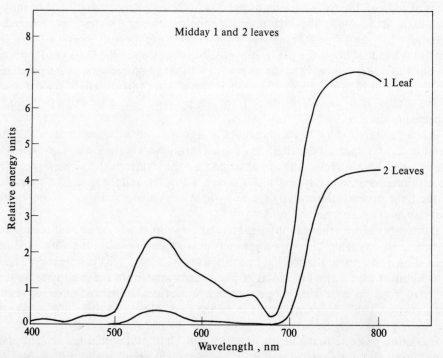

Fig. 7.5 Calculated spectral distribution of midday summer daylight filtered through one or two leaves of sugar-beet. These curves were constructed from the data in Figs. 2.5 and 7.4. Note the high far-red and low red energy levels (from Smith).[402]

150   Phytochrome and photomorphogenesis

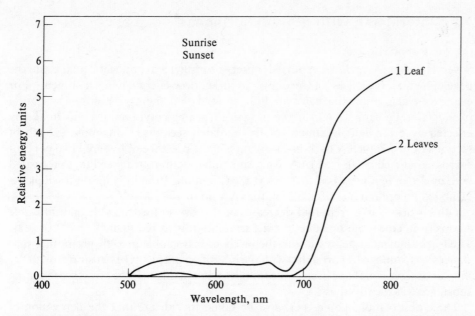

Fig. 7.6 Calculated spectral distribution of summer daylight at sunrise and sunset filtered through one or two leaves of sugar-beet. These curves were constructed from the data in Figs. 2.5 and 7.4. Here there is an even greater ratio of far-red to red energy levels than in Fig. 7.5 (from Smith).[402]

light, leading to very large increases in internode expansion. These effects could be interpreted as a reaction to shading by increased growth, which could enable the plant to reach canopy height, and thus obtain full daylight.

Under more extreme conditions of shading, the high relative far-red content of the light would maintain virtually all the phytochrome in the Pr form, thus preventing the germination of any light-sensitive seed present on the surface of the soil. It is even possible that the germination and growth of woodland ephemerals before the canopy is complete is, at least in part, due to the high Pfr/Ptotal obtaining in early spring.

Table 7.3 Phytochrome photoequilibria under natural conditions.
Avena coleoptiles, dark-grown and minus the leaves, were placed in the cuvette of an Asco Ratiospect R2, exposed either to direct midday sunlight, or to sunlight filtered through one or two thicknesses of sugar-beet leaves, and the Pfr and Ptotal determined. The calculated values were determined by integration of the curves found in Figs. 2.4, 4.11b, and 7.4 (from Smith[402]).

Time of day	No. of leaves	Pfr/Ptotal calculated (per cent)	Pfr/Ptotal observed (per cent)
Midday	0	49·6	53·0–60·0
	1	20·4	17·0–26·0
	2	5·65	5·0–14·0
Sunset or sunrise	0	34·7	—
	1	8·7	—
	2	2·1	—

7.6 Interaction with growth hormones

(a) Auxins and ethylene

Several investigators have reported observations which implicate auxins in the developmental responses of seedlings to brief, low-energy light treatment. For example, Galston and Baker[138] in 1951 showed that the optimum concentration of indole acetic acid (IAA) for the growth of pea epicotyl sections was markedly affected by prior light treatment of the etiolated seedlings. Maximum growth of sections from completely dark-grown plants, in the presence of buffered two per cent sucrose, occurred at 10^{-7} M IAA, whereas similar sections from seedlings pretreated with brief red light grew best at 10^{-5} M IAA. Sections from fully light-grown plants exhibited no optimum concentration for IAA up to 10^{-3} M.[139]

Other workers have reported decreases in the levels of extractable and diffusible auxins from coleoptile tips due to brief irradiation with red light.[24, 48, 443] In peas, a red-light-mediated depression in the levels of extractable growth-promoting substances was found to be correlated with the inhibition of epicotyl extension growth.[325] Similar observations have been made for beans, but in this case the growth-promoting substance was identified as IAA.[119]

These observations could, perhaps, be taken to indicate that the depression of endogenous IAA levels by red light is a causal step in the photomorphogenic processes. In most of the above cases, however, application of IAA has no effect on the red-light-mediated growth effects. For example, with bean seedlings, although red light resulted in a decrease of endogenous auxin levels, exogenous IAA did not relieve the growth inhibition.[119] With pea seedlings, on the other hand, red light only inhibits the so-called 'endogenous growth' of epicotyl sections, and does not affect the extra growth caused by IAA applications.[15, 188] The involvement of auxin in these photomorphogenic processes is therefore not yet fully substantiated.

In another system, however, stronger evidence for a role for auxin has been obtained. The hypocotylar hook of the Black Valentine bean, *Phaseolus vulgaris*, can be caused to open by brief red light in a manner which is consistent with a phytochrome response.[247, 464] This response is a growth response caused by an increased rate of extension of the cells on the concave side of the hook. The opening of the hook, by light, is inhibited by both the presence of the terminal organs, and by the addition of IAA to excised hook segments. These results led W. H. Klein *et al.*[247] in 1956 to suggest that the role of red light *in vivo* was to reduce the supply of auxin from the terminal organs to the hook, thus allowing the differential growth response prevented by auxin. This hypothesis was supported by the observation of R. M. Klein,[241] nine years later, that *p*-chlorophenoxyisobutyric acid (PCIB), an auxin antagonist, caused a slight, but significant, hook opening in the absence of red light. Thus, here again, the hypothesis was proposed that red light causes a reduction in endogenous auxin levels.

An effect of red light on auxin levels could be due to one or more of several possible processes: e.g., effects on auxin production, on auxin degradation, on auxin transport, or on inhibitors or cofactors of auxin action. Attempts to resolve this question have not been conspicuously successful, although some of the findings on possible effects of red light on auxin degradation are of interest.

In 1957 Hillman and Galston[194] observed that the levels of a complex enzyme system in pea seedlings, responsible for the oxidative degradation of IAA *in vitro* (known loosely as IAA-oxidase), were markedly affected by red light treatment. After red light treatment considerably less IAA-oxidase activity was detectable, and the decrease was reversible by far-red light in the classical manner. It was found that the changes in IAA-oxidase levels were due to changes in low molecular weight inhibitors, and cofactors, of the enzyme system, the effective substances being later identified as glycosides and acylated glycosides of the flavonols kaempferol and quercetin.[132, 309] The kaempferol compounds acted as cofactors to the oxidation of IAA by IAA-oxidase preparations *in vitro* when supplied at physiological concentrations, whereas the quercetin components acted as inhibitors of IAA-oxidase at all concentrations.

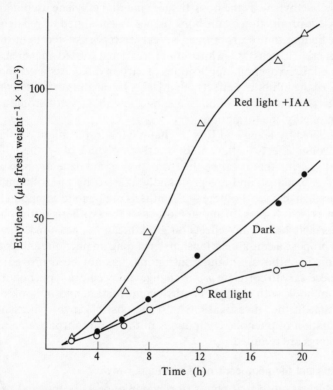

Fig. 7.7 **Effect of red light with and without simultaneous application of auxin (IAA) on the time-course of ethylene production by pea epicotyl hooks (from Kang and Ray).**[225]

Owing to the difficulties of extraction and quantitization of the flavonoids, considerable contradiction exists in the literature as to the precise effects of red light on the levels of the kaempferol and quercetin compounds. It is now known, however, that although red light leads to a decrease in extractable IAA-oxidase activity, and to effects on the levels of potential inhibitors and cofactors of IAA-oxidase, the pattern of changes is too complex for there to exist a simple causal relationship between the two phenomena.[404] Furthermore, it should be pointed out that a decrease in IAA-oxidase activity as estimated *in vitro* would, if reflected in similar changes within the plant, lead to an *increased* level of endogenous auxin, and not to a *decreased* level as found by most investigators. It does not seem possible, therefore, to make a positive correlation between the effects of light on seedling growth, auxin levels, and flavonoid components.

On the other hand, a recent, ingenious, hypothesis has linked the effects of light on bean hypocotyl hook opening with those on polyphenolic substances (such as flavonoids), although effects on auxins have been discounted. Kang and Ray[224-226] followed up the observations of W. H. Klein[247] and Withrow[464] described above, which appeared to provide evidence for a role for auxin in the hook-opening response. They found, however, that light energies just sufficient to cause hook opening had no effect on the amounts of diffusible auxin released by the hook tissue. Furthermore, they showed that applications of exogenous auxin (i.e., IAA and 2,4-D) stimulated the production of ethylene by the hook tissues, and that ethylene itself inhibited hook opening. Ethylene production by the hook tissues was inhibited by red light treatment (Fig. 7.7) – a finding that corroborated an earlier report of red light inhibition of ethylene production in peas[145] – while similar treatment caused an increase in carbon dioxide production. Exogenous applications of carbon dioxide caused hook opening, and antagonized the inhibitory effects of ethylene. It was thus proposed that ethylene was acting *in vivo* as a regulator of hook opening, and that phytochrome acts through the regulation of ethylene formation.

It was considered by Kang and Ray[226] that phytochrome might regulate ethylene production through effects on flavonoid, or other polyphenol synthesis. The known enzymological systems that catalyse the formation of ethylene are basically peroxidases (as is IAA-oxidase), and are profoundly affected by phenolic inhibitors and cofactors. Since light causes the increased synthesis of phenolic compounds, in a wide variety of tissues (see chapter 8), and in most cases the synthesis of the phenolic substances is detectable before any effects on growth can be measured, then it may be reasonable to propose a causal relationship depending on phenolic effects on ethylene formation. This hypothesis, although intriguing, has not been proved, and it will obviously be necessary to correlate *in vivo* changes in specific polyphenolic compounds within the bean hook with changes in ethylene formation and in hook opening.

It is clear that further investigations into the possible involvement of the light-mediated formation of phenolic compounds in the photomorphogenic responses of etiolated seedlings are required.

(b) Gibberellins and the photocontrol of seedling growth

In several species of cultivated plants, for example, peas, beans, and maize, dwarf varieties have been selected out by the plant breeders. Such dwarf varieties are of great economic importance since yield as a proportion of total plant mass is usually greater than with the normal 'tall' varieties. For our purposes, however, the importance of dwarf varieties is in a more fundamental sense.

It is now well established that many dwarf varieties only manifest their dwarf characteristics when grown in the light. Etiolated seedlings of such dwarf varieties thus exhibit a growth rate very similar to those of 'tall' varieties of the same species. In most cases, therefore, the developmental responses of dwarf varieties to light treatments are very much more pronounced than are those of the related tall varieties. A species which has been particularly well investigated is *Pisum sativum*, the garden pea.

When etiolated seedlings of Alaska peas (a tall variety) are placed in low- or high-irradiance light, the growth is retarded, but only temporarily, since within 24 hours the growth rate increases to that found in the fully dark-grown seedlings. Dwarf varieties (e.g., Progress), on the other hand, exhibit a permanent decrease in growth rate upon transfer from darkness to light, and furthermore, the magnitude of the inhibition of growth rate is much greater. In both cases, the light-inhibition of elongation growth can be overcome by applied gibberellins.[149,263,264,266] These results led Lockhart[264] to postulate that light acts through decreasing the rate of gibberellin synthesis within the plant. In the case of dwarf varieties, light would be considered to depress the rate of gibberellin synthesis markedly and permanently, whereas in tall varieties, the depression would be of much lower magnitude and would be temporary.

There are other cases in which applications of exogenous gibberellins prevents the developmental responses to red light. The stimulatory effect of red light on hook opening in the bean, for example, is prevented by gibberellins,[244] and gibberellins also reverse the red light stimulation of flavonoid synthesis in Alaska pea stems.[366] On the other hand, there are many red light-mediated responses which are not reversed by gibberellins. Even in Alaska pea, it has been shown that red light can cause the inhibition of the growth of excised stem sections in the presence of saturating concentrations of gibberellic acid.[188] Mohr and Appuhn[298] could detect no interaction whatsoever between light and gibberellin in the growth responses of that other most popular photomorphogenic test plant, *Sinapis alba*.

From these observations, and several more in the same vein, it would seem unwise to conclude that the function of phytochrome is to cause a drop in gibberellin synthesis, and thus to regulate stem growth. Attempts to detect decreased levels of endogenous gibberellins in light-treated seedlings, of both dwarf and tall pea varieties, have, to date, been unsuccessful. There are now known to be at least 44 chemically different gibberellin substances with varying distributions in plant tissues; these are termed GA_1, GA_2, etc. – gibberellic acid, the first gibberellin to be isolated is known as GA_3. Because of the very large number of potentially active substances present in plants, it is difficult to be certain that all such fractions have been estimated. However, Kende and Lang,[229] for example, have shown that two separable gibberellin fractions occur in extracts of dark-grown and light-treated dwarf peas (*cv.* Progress) and that the levels of the two fractions are not changed by light treatment. One fraction is probably GA_1, while the other is probably GA_5, and Kende and Lang observed that light treatment resulted in a decreased sensitivity of the tissues to the GA_5-like fraction. It thus seems likely that the effects of light are not mediated through effects on endogenous gibberellin synthesis, but are due to a changed sensitivity of the growth processes to gibberellin. On this hypothesis, interaction with gibberellin would be possible only in those cases where the particular light-regulated process under investigation is limited by a gibberellin-dependent step.

Recent evidence has provided support for the conclusion of Kende and Lang that light regulates the sensitivity to gibberellin rather than the synthesis of gibberellin itself. Musgrave *et al.*,[311] working again with Progress peas, have shown that radioactively-labelled GA_1 and GA_5 are accumulated by excised stem sections in accordance with the sensitivity of the sections to growth stimulation by the gibberellins. For example, apical tissues, whose growth responded dramatically to both GA_1 and GA_5, accumulated markedly more radioactive GA_1 and GA_5, than did basal tissues which were largely insensitive to gibberellin. Chemical analogues of gibberellic acid with very little growth promoting activity were not taken up preferentially by the apical tissues. Light treatment, which as stated above, reduces the growth responsivity of the tissues to gibberellin, particularly to GA_5, also reduced the binding of the labelled GA_5 to the tissue. The evidence suggests that the accumulation is probably due to the binding of the hormones to specific gibberellin receptors.

The authors suggest that light may act in one of two ways in this response: that it may alter the binding characteristics of the gibberellin receptors, thus lowering the affinity for gibberellins; or it may inhibit the synthesis of receptor molecules, thus reducing the number of binding sites. In view of the observed 3 hour lag before the effect of light on gibberellin accumulation becomes detectable, the second possibility was considered most likely.

At the present time therefore the most logical conclusion seems to be that light may cause a depression in the sensitivity of stem tissue to gibberellin, and that this may be due to effects on the formation of specific gibberellin receptors. In cases where overall growth is limited by a gibberellin-dependent step, then application of high concentrations of exogenous gibberellin would reverse the effects of light treatment.

The above generalization probably holds true only for stem growth, since there is compelling evidence that light treatment of the leaves of certain etiolated plants causes a dramatic, and rapid, surge in the levels of endogenous gibberellin. As has been described in earlier sections of this chapter, the growth rate of many leaves is increased by red light in a characteristic phytochrome-regulated manner. This is most easily observed in dicotyledonous plants, but an analogous phenomenon occurs in the leaves of cereals. When cereal seedlings are grown in total darkness, their leaves remain tightly rolled around a longitudinal axis. When the seedlings are transferred to daylight, the leaves unroll by a mechanism involving expansion of the cells of the upper mesophyll region of the leaves. This unrolling phenomenon can be studied in isolated sections of etiolated leaves, floating on the surface of water, in a petri dish. Virgin[448] in 1962 and Klein et al.,[245] in 1963, showed that the unrolling phenomenon was a highly sensitive phytochrome response.

In 1968, Reid et al.[352] detected a very rapid transient increase in endogenous gibberellin levels, in sections of etiolated barley leaves, upon irradiation with red light. This work was followed up by Wareing and his colleagues,[12] who found that the increase in gibberellin levels was extremely rapid reaching a maximum at approximately 15 minutes after the start of the red light treatment, and followed by an equally rapid decline (Fig. 7.8). The increase in gibberellin content in response to red light treatment could be prevented by pretreatment of the sections with 2-chloroethyl trimethylammonium chloride (CCC), an inhibitor of gibberellin biosynthesis,[352] and it was concluded that a rapid, phytochrome-mediated stimulation of gibberellin synthesis was responsible for the rise in gibberellin levels. However, it seems unlikely that such rapid transient changes in gibberellin level could be due to synthesis, and release of gibberellin from a bound form followed by its re-complexing back into the presumably inactive bound form, is now considered most probable.[12]

Interactions have also been observed with abscisic acid (ABA). If sections are pre-incubated with ABA before red light treatment, both the growth responses, and the gibberellin increases, are lost.[12] This is not a simple case of ABA-gibberellic acid interaction, however, since abscisic acid treatment is still inhibitory to unrolling when applied 4 hours after light treatment, long after the transient rise in gibberellin levels has occurred.

It is, however, difficult to determine whether or not the undoubted surges in

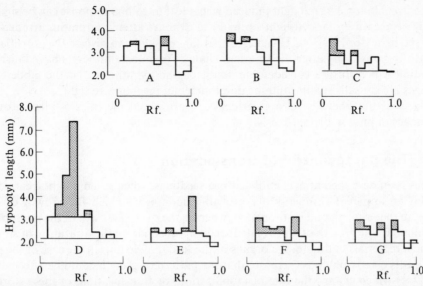

Fig. 7.8 Effect of red light on endogenous gibberellin levels in etiolated wheat leaves. Each separate histogram represents the gibberellin activity present in extracts of leaves after separation into ten fractions by paper chromatography. Each fraction was tested by its promoting effect on the growth of lettuce hypocotyls. The solid parts of the histograms represent gibberellin activity significantly above control.

A = dark-grown leaves
B = 5 min red light
C = 5 min red light + 5 min dark
D = 5 min red light + 10 min dark
E = 5 min red light + 20 min dark
F = 5 min red light + 30 min dark
G = 5 min red light + 60 min dark

(from Beevers *et al.*).[12]

giberellin really are related to the ultimate developmental response of unrolling. It is quite certain that application of exogenous gibberellin to dark-grown leaf segments causes unrolling, but it is equally true that exogenous cytokinins also cause unrolling (see Table 7.4), and it is thus not possible to impute any specific role for the gibberellins. Furthermore, it is not yet known whether the radiant energy relationships for the induction of gibberellin increases are similar to those for the induction of the unrolling phenomenon – the unrolling phenomenon in wheat leaves is saturated at very low light levels (Fig. 3.3) and it would be interesting to know whether or not such low radiant energies cause increases in gibberellin levels. A further point of difficulty arises from a consideration of the time-course of the escape from

Table 7.4 Effects of gibberellic acid and kinetin on unrolling of wheat leaf segments in the dark (calculated from data in Beevers *et al.*[12]).

	Final leaf width (mm) in the stated hormone concentration (mg/l)				
	0	2	5	10	20
Gibberellic acid	1·59	1·98	2·10	1·96	—
Kinetin	1·29	1·43	1·62	1·73	1·59

photoreversibility. The red light induced wheat leaf unrolling response can be significantly decreased by far-red light up to $1\frac{1}{2}$ to 2 hours after a saturating irradiation with red light (see Fig. 3.4). This means that far-red light can reverse the potentiated growth response long after the transient changes in gibberellin levels have finished, a finding which throws considerable doubt on the hypothesis that the gibberellin changes are causally involved in the developmental responses to light.

It is clear that there is yet a great deal to learn about the roles of plant growth hormones in photomorphogenesis.

7.7 The photocontrol of translocation

As has been described above, etiolated pea seedlings, when given a light treatment, exhibit an apparent compensatory growth response, in which the growth of the leaves in the terminal bud is stimulated, whereas the overall growth of the internodes is inhibited. The growth of all etiolated seedlings depends on the utilization of stored nutrient materials in the storage organs of the seed (cotyledons in the case of the pea seed). It is clearly obvious therefore that the pattern of growth could be affected by processes which change the rate of mobilization, or translocation, of these storage materials. Evidence that photomorphogenic phenomena may involve effects on translocation has been accumulating in recent years.

If the roots and cotyledons of etiolated pea seedlings are removed, growth will proceed in darkness if the cut ends of the epicotyls are kept in water. Light treatment under these conditions, however, does not lead to a stimulation of leaf growth. If the water is replaced by a sucrose solution, on the other hand, the responses to light are fully restored, indicating that the uptake of carbohydrate nutrients is essential for the photomorphogenic processes. Furthermore, it has been shown that brief red light treatment of the isolated pea shoots leads to dramatic increases in the incorporation of basally-fed ^{14}C-sucrose into the terminal buds. These effects are red/far-red reversible, and thus are classical phytochrome responses. Goren and Galston[148] have shown that the effect of red light on sucrose translocation into the terminal buds is apparent within one hour of light treatment, whereas the changes in terminal bud growth do not become measurable until 3–4 hours after treatment (Fig. 7.9).

Light effects on translocation are not restricted to sucrose. Although sucrose translocation is most sensitive to red light, the transport of maltose and fructose, but surprisingly not glucose, is also under phytochrome control.[148] Furthermore, other workers have shown that the translocation of certain inorganic ions (e.g. potassium, rubidium and calcium[248]) and of intermediates of flavonoid synthesis (i.e., phenylalanine and cinnamic acid;)[166] is also markedly affected by the photoactivation of phytochrome.

These results demonstrate that it is not always sensible to search for the mechanism of phytochrome action within the tissue, or organ, which exhibits the ultimate developmental responses. The interrelationships between different parts of the plants, as manifested in the transport of important metabolites (be they nutrients, or hormones) between such interacting tissues, is of fundamental importance to the final pattern of development. Many cases are known in which isolation of an organ, from the neighbouring tissues, destroys the developmental sensitivity of the excised organ to light. The etiolated pea seedling is again a very good example of this generalization. When terminal buds are excised from etiolated seedlings, and floated on a two per cent sucrose solution, a certain amount of growth occurs in darkness, but the normal light-stimulation of growth observed in intact seedlings is not detectable. If, however, a portion of the epicotyl is excised with the terminal bud, some light-stimulation of bud growth is detectable, and the degree of stimulation is directly proportional to the length of stem left attached to the bud.[148] This effect is obviously not an effect on sucrose uptake since sucrose at saturating concentration is available in all cases. It seems that the presence of epicotyl tissue is essential for the light-growth response of the terminal buds.

It appears, therefore, that the photomorphogenic phenomena of etiolated seedlings involve a large number of separate, but interrelated, physiological processes. Full understanding is unlikely to be achieved without a great increase in our knowledge of the correlation of developmental activities in the plant. Attempts to understand the mechanisms whereby light regulates development are currently centred on the molecular, and biochemical, events intervening between the initial perception of the light, and the ultimate morphogenic changes. The next, and penultimate, chapter of this book deals with the present state of knowledge in these areas; it should be borne in mind, however, that as yet we have very little conception of how development is regulated in any organism, and thus the available information on photomorphogenesis is fragmentary and preliminary, and will remain largely so until an acceptable general theory of development appears.

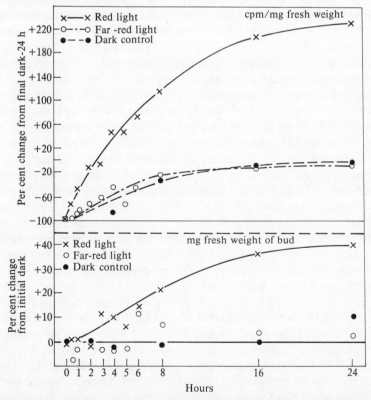

Fig. 7.9 Effect of brief red light, with and without subsequent brief far-red light, on the growth in weight of pea terminal buds (lower curves), and on the incorporation of ^{14}C-sucrose into the buds (upper curves) (from Goren and Galston).[148]

8

The biochemistry of photomorphogenesis

It is quite clear from much of the foregoing that the primary action of phytochrome is exerted very rapidly, probably through effects on the permeability, or other properties, of certain cellular membranes. As far as rapidity of action is concerned, the same is probably true of the blue absorbing photoreceptor, although its continued action appears to require continuous irradiation.

At the other end of the timescale, the morphogenic effects of light do not normally become detectable for hours, or even days, after the onset of the light treatment. Between the primary act of the photoreceptor and the ultimate developmental phenomena, a large number of partial processes must be involved which should manifest themselves as biochemical changes. We might reasonably expect, for example, to observe changes in the transcription and translation of messenger-RNA, changes in the amounts and types of enzyme synthesized, changes in the activity of existing enzymes, or changes in the rates of specific enzyme turnover.

The concentration of our thoughts on enzymes reflects the universal view that differentiational changes must ultimately be controlled through the regulation of gene expression, leading through changed enzyme complements to the observed changes in cell form and function. It is not meant to imply that phytochrome operates directly on gene expression – as already stated this seems most unlikely – but rather that the consequential result of the primary action of phytochrome is the regulation of gene expression, leading to a changed enzyme complement within the cell.

If this is true, we are faced with the problem of determining whether the photomorphogenic reaction systems operate by controlling transcription, translation, enzyme activation and inactivation, or enzyme degradation. This is a daunting task, since it is quite possible that phytochrome, acting through a common primary reaction mechanism in all cells, may, nevertheless, regulate the various developmental processes through different ways. Furthermore, methods for the study of the regulation of gene expression in higher plants are in their infancy. It should therefore not be surprising that the very large number of investigations into the effects of light on plant metabolism, that have been reported in recent years, have been singularly unhelpful to our aim of understanding photomorphogenesis. Many investigations, indeed most, have been purely descriptive, reporting, for example, that a particular enzyme or a particular metabolite is affected by light, without delving more deeply into the biochemical mechanisms responsible for the changes. Such studies may eventually be built up into an overall picture of the metabolic effects of light, but if they are not incisively analytical they are of little use in a consideration of mechanisms.

In order to cover in a logical manner the many diverse and apparently unrelated investigations into the biochemistry of photomorphogenesis, the various findings will be described here more or less in order of their proximity to the initial photoact. This treatment gives the appearance of a temporal sequence of events but the reader should not readily assume that a causal relationship holds. It is certainly possible that the earlier known events in photomorphogenesis are prerequisites for the later events, but as yet very few attempts have been made to follow through the biochemical changes brought about by light in any one plant. This chapter ends with an attempt to bring together the fragmentary information that is available for a preliminary understanding of the control mechanisms operating between the perception of the light stimulus, and the ultimate developmental changes.

8.1　Rapid effects related to membrane changes

(a)　Bio-electric potentials

The earliest known event to succeed the perception of light by phytochrome is the change in electric potential of the etiolated oat coleoptile. As briefly mentioned above (page 111), Newman and Briggs,[315] using a sensitive flowing-drop technique (which obviates the problems associated with direct contact between the electrode and the plant tissue), showed that red light caused an increase in the positive potential existing between the upper part and the base of the coleoptile. The lag in this response is less than 15 seconds (Fig. 8.1) and the increases are of the order of 5–10 mV, which enables them to be reliably measured. In subsequent darkness, the potential returns to the original level within a few minutes, after which far-red light can be shown to cause a significant decrease, again with a lag of less than 15 seconds. Clearly, therefore, this extremely rapid response is phytochrome-mediated. Furthermore, since maximum response was only observed when the whole coleoptile was irradiated, the authors concluded that the changes in potential represent an integration of the responses of many cells. In addition, it was found that blue light caused a massive decrease in potential of up to 40 mV using pea epicotyls, but since this interesting response was apparently unrelated to phytochrome, it was not investigated further.

Other phytochrome-mediated effects on bio-electric potentials are known. Jaffe, who was the first to observe such effects, found that red light caused the apex of mung bean root sections to become more electropositive, with a lag of less than 30 seconds.[210] The increases in potential were considerably smaller (ca. 1 mV) than those later observed by Newman and Briggs for the coleoptile, but this may have been partly due to the use of different techniques. Red/far-red reversibility was observed, thus showing again that phytochrome is the photoreceptor.

(b)　Root tip adhesion – the 'Tanada effect'

The above observations of extremely rapid changes in bio-electric potentials can be directly related to (indeed they stem from), earlier findings of rapid events best explained in terms of membrane changes, i.e., root tip adhesion and nyctinastic leaf movements. The adhesion of barley[416] and mung bean[415] root tips to a negatively charged glass vessel (the 'Tanada effect') occurs with a time course similar to that of the light-mediated bio-electric potential changes in root tips. The response is quite remarkable and one marvels at the way in which it must first have been discovered; the root tips of barley will only adhere to the glass in the presence of precise

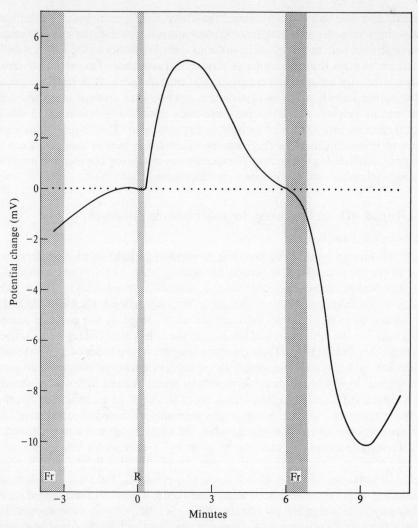

Fig. 8.1 The time-course of the changes in electrical potential of the oat coleoptile upon irradiation with either 10 seconds of red or 65 seconds of far-red light. The data represent the average of five coleoptiles (after Newman and Briggs).[315]

concentrations of ATP, indole acetic acid, L-ascorbic acid, manganese chloride, potassium chloride and hydrochloride acid, while mung bean roots also require calcium ions! In addition to this complex medium, the glass vessel must be scrupulously cleaned with a non-ionic detergent, followed by rinsing in orthophosphate solution, in order to negatively charge the glass surface. If this is not enough, optimum results can only be obtained with the tips of secondary roots of seedlings grown in deionized water.[470]

When all these conditions are fulfilled, root tips will normally not adhere to the surface under dim green safelights, but upon irradiation with red light, a large proportion of the roots adhere to the surface by the apices, the first response being detectable within about 30 seconds. Subsequently, the root tips can be released by treatment with far-red light, again showing the involvement of phytochrome. The

'Tanada effect' has recently been placed on a technically more controllable level by the use of electrodes in the bathing solution to which a potential difference is applied.[348] The root tip then attaches to the cathode. In this way the voltage can be varied such that the point at which the adhered roots are released can be determined, enabling a more acceptable quantitative measure of adhesion to be made.

Phytochrome is, apparently, not the only agency which can control the adhesion of the root tips; Tanada[417] has recently shown, for example, that abscisic acid (ABA) will cause adhesion of the root tips in darkness, while supra-optimal concentrations of indole acetic acid (IAA) prevent adhesion, even in the presence of red light. Indole acetic acid and ABA apparently act antagonistically in this response.[419,420]

Another substance which can regulate the adhesion of root tips is the mammalian neurohumour acetylcholine (ACh). Jaffe,[211] using a specific clam-heart bioassay, showed that ACh was apparently present in mung bean root tips and, moreover, red light caused an efflux of ACh from the tips, and increased the internal levels within 4 minutes of irradiation. Subsequent far-red light reduced the internal levels to those of the control. Added ACh substituted for red light in several responses, including the induction of root tip adhesion, the stimulation of hydrogen ion efflux, and the inhibition of secondary root initiation. An inhibitor of ACh action in mammals, atropine, and an ACh degrading enzyme, acetylcholinesterase, both reduced the red light-mediated induction of root tip adhesion when present in the bathing medium; on the other hand, eserine, an inhibitor of acetylcholinesterase, inhibited the far-red light-induced release of the root tips.

Although ACh has been found to cause sporulation in the fungus *Trichoderma*[155] (normally a light-initiated process, but not via phytochrome), there is little evidence for a role for ACh in other phytochrome controlled phenomena. Mohr and co-workers,[228] indeed, found that ACh did not substitute for continuous far-red light in the photomorphogenesis of *Sinapis alba* even in the presence of eserine to inhibit endogenous acetylcholinesterase. In the case of *Trichoderma* cited above, no response to ACh was obtained unless eserine, the inhibitor of acetylcholinesterase, was also present.

A further, more cogent criticism, comes from Tanada's[418] observation that adhesion due to ACh can be prevented by increasing the potassium ion concentration to 10^{-3} M. Thus, Tanada suggests that ACh may be competing with potassium ions for some site on the plasmalemma and thereby preventing the detachment process, for which the potassium ion is essential. Similar competitive effects can be observed between sodium and potassium ions. It therefore seems unlikely that ACh is acting hormonally to affect membrane permeability in a manner analogous to its effects in mammalian nerve cells, but rather is acting unspecifically as a competing cation. However, as Tanada points out, this in no way invalidates Jaffe's finding of changed endogenous levels of ACh in the root upon light treatment. These changes are so rapid, and of such magnitude, that acetylcholine must be a prime candidate for an immediate metabolite of phytochrome action, and as such merits intensive investigation in a range of photomorphogenic responses.

(c) Nyctinastic leaflet closure

The closure of the leaves of many leguminous plants when removed from light (known as nyctinasty, or sleep-movement) is the other well-studied phenomenon which has led to the view that phytochrome regulates membrane permeability. In

1966 Fondeville *et al.*[122] showed that the normal closure of *Mimosa pudica* leaflets, upon transfer from light to dark conditions, could be inhibited by a brief irradiation with far-red light. Subsequent red light negated the effect of far-red, and thus the phenomenon was seen to be under phytochrome control. Similar phenomena have been found in a range of other legumes[195] and have been particularly well studied in *Albizzia julibrissin.*[213]

The closing movement is due to changes in the volumes of the sub-epidermal cells – the so-called 'motor cells' – of the pulvinules. Figure 8.2 shows how these volume changes lead to movements in the leaflets. In Fig. 8.3, the time-course of leaflet closure in *Albizzia julibrissin* is shown, together with the effects of red and far-red light. It can be seen that phytochrome acts in this response within about 10 minutes.

With *Albizzia julibrissin*, leaflet pairs can be removed from the plant and floated on water in a petri dish. In this way the closing response can be readily investigated.

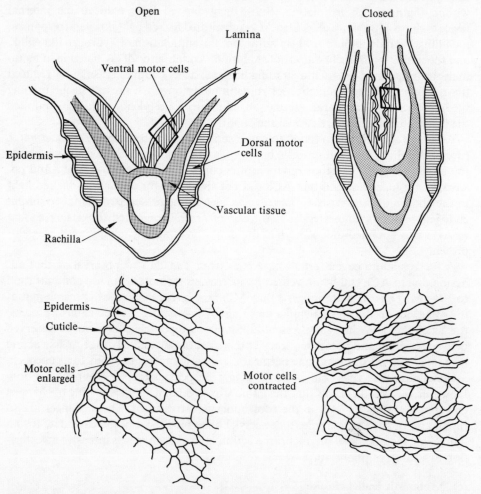

Fig. 8.2 The anatomy of the leaflet closing reaction in *Albizzia julibrissin*. *Upper left*, diagrammatic transverse section across the pulvinal region of an open leaflet pair; *upper right*, the same pair when closed; in both cases the portion within the rectangle is shown magnified below to illustrate the changes in the size of the ventral motor cells. (Redrawn from photomicrographs; from Satter *et al.*)[372]

164 Phytochrome and photomorphogenesis

The first observation with this experimental system was that irradiation with red light caused a significant leakage of electrolytes from the severed ends of the rachilla into the medium in the petri dish.[213] Later Satter et al.[371] were able to show that red light promotes and far-red inhibits the efflux of potassium ions from the ventral motor cells; on the other hand, the movement of potassium ions *into* the dorsal motor cells occurs in darkness, and is not affected significantly by red or far-red light. These impressive results were obtained by the use of a powerful new technique, electron microprobe analysis. In this instrument, an electron beam is focused onto a section of tissue, and this beam excites atoms to emit X-rays whose frequency is characteristic of the element. The element-specific X-rays are gathered, analysed and quantitized, thus providing information on the identity and amounts of particular elements present in the region of tissue in the electron beam. Satter and Galston have since used this sensitive method to show that potassium ion changes correlate not only with phytochrome-mediated leaflet movements, but also with movements regulated by endogenous rhythms.[369, 370]

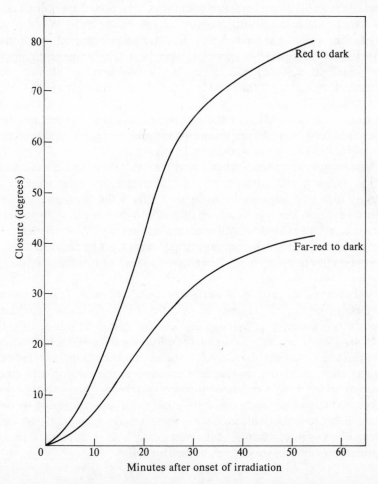

Fig. 8.3 The time-course of leaflet closing in *Albizzia julibrissin* upon transfer from continuous white light to darkness, preceded by either red light (red to dark) or far-red light (far-red to dark) (after Satter *et al.*).[371]

These studies have allowed the construction of a hypothesis accounting for the phytochrome control of leaflet closure[371]:

1. Water moves into and out of the motor cells in response to changes in osmotic concentration, especially in changes of potassium ions;
2. The efflux of potassium ions through the membranes of the ventral cells requires Pfr, although the movement of potassium ions into the dorsal cells is not directly controlled by phytochrome;
3. Phytochrome is situated in the membrane of the ventral cells adjacent to, or at, the sites through which potassium ions move;
4. The movement of potassium ions across the membrane involves the expenditure of energy, probably in the form of ATP, which may actuate a potassium ion pump. (This last point refers to inhibition of leaflet closure by anaerobic conditions, sodium nitrite and dinitrophenol, all of which are known to impair the respiratory formation of ATP.)

Analysis of the situation in *Mimosa pudica* by Setty and Jaffe[385] showed that the movement of the leaflets was correlated with the contraction of specialized contractile vacuoles found within the central vacuole of the motor cells. When plants were transferred from white light to darkness, leaflet closure occurred, accompanied by the contraction of the contractile vacuoles; when far-red light was given, immediately after the white light, neither leaflet closure nor vacuole contraction occurred in the subsequent darkness. Far-red followed by red light, however, gave leaflet closure and vacuole contraction. Irradiation of a small part of the pulvinus with a microbeam of red light caused the contraction of the contractile vacuoles in other, non-irradiated parts. These results suggest that an intracellular messenger is released from irradiated cells that stimulates all the motor cells in the pulvinus.

How these interesting findings can be integrated with the potassium ion movements reported by Galston and coworkers is not immediately apparent. One obvious possibility is that phytochrome in some way affects the properties of important cellular membranes as a very early step in its action. Whether phytochrome is actually a membrane component, and thus affects membranes directly, or whether it operates to release a membrane-active second messenger cannot yet be decided, but investigations centred on this question might be expected to lead to significant advances in the near future.

Yet another word of caution is necessary here, however. Transmission of the stimulus throughout the cell, as reported by Jaffe, is in accordance with the action of phytochrome being exerted at membranes, and involving the release of active substances. In *Mougeotia*, on the other hand, the effect of a microbeam is not transmitted within the cell to any noticeable extent.[174] Such a state of affairs is more compatible with a direct effect on the microfilaments involved in the chloroplast rotation,[449a] rather than an indirect effect mediated through membrane changes, since one might expect the products of phytochrome-mediated membrane transport to be mobile within the cell and to induce chloroplast movement in the non-irradiated area. Thus, although an effect of phytochrome at the membrane level is currently a very popular concept, other possibilities should not be overlooked.

8.2   Rapid effects not obviously associated with membranes

(a) Gibberellin changes in cereal leaves

When dark-grown cereal seedlings are irradiated with red light, a surge of gibberellin can be detected in the leaves[12, 269, 351, 352] (see page 156). These changes occur within 10–20 minutes (see Fig. 7.6), and have been considered to be due to the phytochrome-mediated release of a gibberellin from a bound form.[269] This phenomenon is discussed in some detail in chapter 7 where it is considered in relation to interactions between phytochrome and hormones. It is, however, also very interesting from the standpoint of the biochemistry of phytochrome action since it has recently been found that irradiation of a crude homogenate of barley leaves with red light leads to an increase in detectable gibberellin levels within about 16 minutes.[353] Such homogenates, furthermore, are apparently capable of converting added radioactive gibberellin A_9 (GA_9) to other gibberellins, and this reaction is considerably stimulated by red light.

Plant materials can contain a large number of gibberellin-like substances and it appears that interconversion of different gibberellins may be an important aspect of the synthesis of specific gibberellins. It is possible therefore that red light might act in the intact leaves to increase the rates of the later steps of gibberellin biosynthesis rather than to cause release of gibberellin from a bound form. Furthermore, it is significant that these changes are detectable in leaf homogenates proving that enzyme synthesis cannot possibly be involved in the response to red light.

Unfortunately, the identity of the gibberellins formed from red light in the homogenate was not determined in this investigation, nor was the involvement of phytochrome rigorously proven by photoreversibility experiments. Nevertheless, the phenomenon is clearly worthy of intensive investigation since it represents one of the very few cases in which phytochrome appears to be acting in a cell free system.

(b) Changes in adenosine triphosphate and adenosine diphosphate

Recently, there have been two reports of very rapid changes in the concentrations of ATP and ADP in plant tissues. In secondary root tips of mung bean, 4 minutes of exposure to red light markedly decreased the levels of ATP in the tissues with no comparable effects on ADP.[471] In *Avena* mesocotyl tissue, on the other hand, red light caused a marked increase in ATP concentration and a concomitant, and almost stoichiometric, drop in ADP concentration.[368] In the latter case, the changes were complete within one minute and considerable action had occurred within 30 seconds (Fig. 8.4); in addition the concentrations of ATP and ADP returned to the original dark level within 5 minutes of switching off the red light. Consequently, it was not possible to carry out classical red/far-red photoreversibility experiments to determine whether phytochrome was the photoreceptor. The low-energy requirement, however, is consistent with the involvement of phytochrome. In the case of red light-mediated depression of ATP concentration in mung bean roots, red/far-red photoreversibility experiments confirmed that phytochrome was the photoreceptor.

In these two tissues, therefore, red light leads to changes in ATP levels, but in one case an increase is observed and in the other a decrease. In *Avena* mesocotyl sections, however, the immediate and transient increase in ATP levels is followed, 1–2 hours after irradiation, by a fall relative to the dark controls. No changes are seen in ADP levels. Thus, it is possible that these apparently contradictory responses to red light in two tissues may be due to different time-courses of a single process.

In any case, the effects of light on ATP and ADP could with benefit be investigated in a range of photomorphogenic test materials, since it is easy to imagine that such changes could have far-reaching effects on cell metabolism.

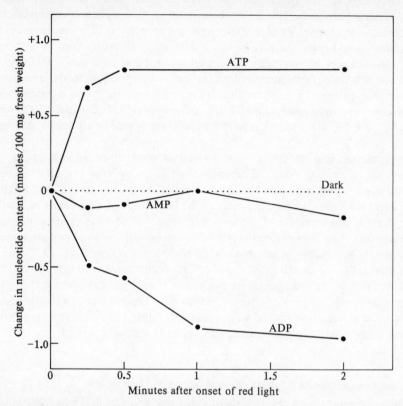

Fig. 8.4 The time-course of ATP, ADP and AMP changes in etiolated oat mesocotyl tissue upon irradiation with red light (reprinted with permission from Sandmeier and Ivart).[368]

(c) Phytochrome and nicotinamide adenine dinucleotide (NAD)-kinase

Effects of phytochrome on the levels of other nucleotide coenzymes have also been reported. Tezuka and Yamamoto,[423] in 1969, reported that a brief irradiation with red light, given in the middle of the night to light-grown *Pharbitis nil* seedlings, caused the tissue concentrations of nicotinamide adenine dinucleotide phosphate (NADP) to increase. The most significant feature of these experiments was the finding that a partially purified phytochrome preparation from etiolated peas had NAD-kinase activity which could be stimulated *in vitro* by red light. NAD-kinase is the enzyme which catalyses the phosphorylation of NAD to NADP in the following reaction:

$$NAD + ATP \xrightarrow{\text{NAD-kinase}} NADP + ADP$$

The effect of red light on the enzyme was to lower its Michaelis constant (K_m) for NAD. In simple terms this is equivalent to increasing the affinity of the enzyme for

168   Phytochrome and photomorphogenesis

the NAD substrate thereby increasing the capacity of the enzyme to catalyse the reaction. The effect of red light on the K_m could be reversed by far-red light, and thus a direct interaction between the phytochrome in the preparation and the NAD-kinase was postulated. It was later shown that phytochrome and the NAD-kinase activity could be separated by chromatography on calcium phosphate gels, thereby proving that phytochrome did not have NAD-kinase activity.[424]

These effects on NADP and NAD-kinase have received little support from other workers in the field, and one group has reported a lack of success in repeating the light effect on NAD-kinase. It would seem important, therefore, for other investigators to attempt to confirm these findings since they clearly could form the basis of an attractive hypothesis.[290]

8.3 Enzyme changes – general aspects

(a) The range of enzymes controlled by phytochrome

Very many investigations into the effect of light on enzyme activities have been carried out over recent years. The hope behind this type of work is that information will be obtained of direct value in understanding the photocontrol of seedling development. Unfortunately, very few of the investigations have been carried through to the point at which unequivocal information can be obtained. Much of the work has been of a descriptive nature only with no attempt to determine the mechanisms underlying the observed enzyme changes.

A more or less comprehensive list of those enzymes whose extractable activities are known to be under phytochrome control is given in Table 8.1. A wide range of metabolic activities including photosynthesis, photorespiration, chlorophyll synthesis, fat degradation, starch degradation, nitrate assimilation, nucleic acid synthesis and degradation, and secondary product synthesis may thus be under phytochrome control exerted at the level of specific enzyme activities. Table 8.1 also gives information on the approximate lag phases which intercede between the onset of the light stimulus and the first detectable changes in extractable enzyme activities. Clearly, most of the reported enzyme changes occur only after a considerable period of time, and thus can only be secondary effects of the primary action of phytochrome. We are not likely therefore to derive much of value concerning phytochrome action from a detailed study of such enzymes.

With enzymes which change over a shorter time-course, however, a more detailed analysis may provide useful information on the primary action of the photoreceptor, even though here again, it is unlikely that the photoreceptor *directly* modifies the level of these enzymes. The only enzymes with a sufficiently short lag phase (excluding NAD-kinase which has already been discussed) are lipoxygenase, inorganic pyrophosphatase and phenylalanine ammonia-lyase. Of these only lipoxygenase and phenylalanine ammonia-lyase (PAL) have been investigated in any real detail. Lipoxygenase and PAL are of especial interest since phytochrome apparently acts to depress the activity of the one (lipoxygenase) while it increases the activity of the other (PAL).

(b) Problems of expressing enzyme results

Before considering lipoxygenase and PAL in detail, a few general remarks on the problems associated with enzyme studies are needed.

Table 8.1 Phytochrome-mediated changes in enzyme activity.

Enzyme	Plant	Effect	Lag	Reference
Intermediary metabolism				
NAD-kinase (*in vitro*)	Pea	+	0	423
lipoxygenase	Mustard	–	0	320
amylase	Mustard	+	~6 h	94
ascorbic acid oxidase	Mustard	+	~3 h	93
galactosyl transferase	Mustard	+	~12 h	441
NAD†-linked glyceraldehyde-3-phosphate dehydrogenase	Bean	+	<12 h	117
nitrate-reductase	Pea	+	<2 h	216
Nucleic acid and protein metabolism				
RNA-polymerase (nuclear)	Pea	+	4 h	42
ribonuclease	Lupin	+	4 h	1
amino acid activating enzymes	Pea	+	—	187
Photosynthesis and chlorophyll synthesis				
ribulose-1,5-diphosphate carboxylase	Bean	+	<24 h	117
transketolase	Rye	+	—	116
NADP†-linked glyceraldehyde phosphate dehydrogenase	Bean	+	<12 h	117
alkaline fructose 1,6-diphosphatase	Pea	+	—	150
inorganic pyrophosphatase	Maize	+	~2 h	60
adenylate-kinase	Maize	+	—	60
succinyl CoA-synthetase	Bean	+	—	410
Peroxisome and glyoxisome enzymes				
peroxidase	Mustard	+	72 h	381
glycollate-oxidase	Mustard	+	6 h	444
glyoxylate-reductase	Mustard	+	6 h	444
isocitrate-lyase	Mustard	0	—	227
catalase	Mustard	0	—	95
Secondary product synthesis				
phenylalanine ammonia-lyase	Pea (many others)	+	1·5 h	403
cinnamate-hydroxylase	Pea	+	—	364

First of all, it is always difficult to extract and assay enzymes in such a way that a meaningful estimate of the *in vivo* activity of the enzyme can be obtained. Enzymes are extremely delicate and subject to various types of damage, particularly during extraction and storage. In addition, there is no universally accepted way of expressing results of enzyme determinations such that only *bona fide* changes are shown. Many plant physiologists express results in terms of 'enzyme units per seedling', or 'enzyme units per hypocotyl', etc. On this basis, any environmental change that leads to an increase in overall protein synthesis in the seedling, or organ, will yield an apparent increase in the enzyme under study. Biochemists, in general, disapprove of this way of expressing results and assert that for each enzyme assay, the total protein content of the extract should be determined, and the results expressed as 'enzyme units per unit protein', otherwise known as 'specific activity'. This method has the valuable advantage that changes in overall protein levels are not mistaken for specific changes in a particular enzyme.

The specific activity method should be used in most cases in preference to the 'enzyme unit per seedling' method, but there are occasions when a third method is even better. It is arguable that what is important within the plant is not the total enzyme activity per organ, nor the activity of an enzyme as a proportion of the total protein, but the amount of that enzyme present in each cell. On this basis, 'enzyme units per cell' would be the best method of expressing results. Since this involves the tedious and relatively inaccurate process of estimating the number of cells in each sample of tissue from which an enzyme extract is to be made, it is hardly surprising that this way of expressing the results is very rarely found. Nevertheless, its use is to be encouraged.

(c) Enzyme synthesis, degradation, activation, and inactivation

The possible processes that can contribute to a change in the amount of any particular enzyme within a cell are enzyme synthesis, enzyme degradation, enzyme activation, and enzyme inactivation. The first two processes obviously result in the formation of new enzyme molecules, or the breakdown of existing enzyme molecules, and thus require the activity of complex metabolic sequences. Enzyme activation and inactivation, on the other hand, are processes which result in the *modification* of pre-existing enzyme molecules to make them catalytically respectively more or less active. These modifications may involve covalent changes in the enzyme molecules (e.g., phosphorylation, or release of the enzyme from a bound state), or non-covalent allosteric interactions effected by small metabolites. Clearly, when investigating a light-mediated change in the extractable activity of an enzyme, it is important to determine which one, or more, of these four possible processes is responsible for the change. As will be stressed later, only when a change in the rate of *de novo* synthesis of an enzyme is proven, can it be considered possible that the light stimulus is operating through the regulation of gene expression at the transcription or translation level.

It is not easy to determine whether or not enzyme synthesis, degradation, activation, or inactivation is occurring in respect of any specific enzyme. The level of an enzyme in a cell may be determined at any one point in time by any, or all, of these processes. Figure 8.5 shows schematically how both rises, and falls, in the amount of an enzyme can be caused by the appropriate *relative* changes in the synthesis and degradation rates. The steady-state levels of the enzyme represented by A, C, and E are all very different, but are maintained by identical rates of synthesis and degradation in each case. The steady-state G, on the other hand, is maintained in the complete absence of both synthesis and degradation. The rapid increase seen in B is due to a sudden burst of synthesis, whereas that in F is due to a sudden cessation of degradation. The slow fall in D, on the other hand, is due to a small increase in the rate of degradation. In addition, control of the effective activity of the enzyme molecules in the cell may at any time be superimposed by inactivation as at H, or activation as at J.

In the light of these considerations, it is obvious that any attempt to determine exactly what is happening under a certain set of conditions will be fraught with many difficulties. It should therefore not be surprising that it has not yet been possible to prove, for any phytochrome-mediated increase in enzyme activity, whether the increase is due to an increased rate of synthesis, a decreased rate of degradation, to some modification of pre-existing enzymes, or to a combination of some or all of these processes. On the credit side, however, it is now possible to test for the occurrence of enzyme synthesis, although it is still not possible to estimate the actual *rate* of synthesis.

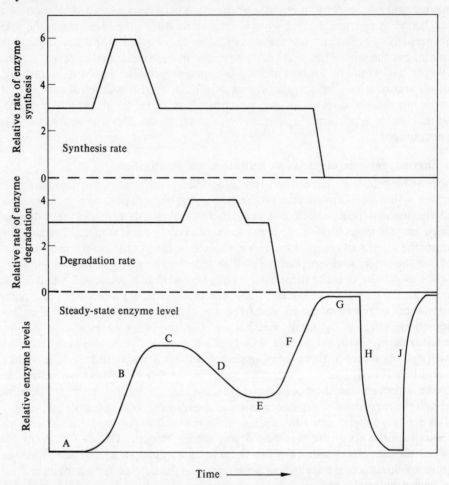

Fig. 8.5 Diagram of the effects of changing synthesis rate, degradation rate, and of activation and inactivation, on the steady-state level of an enzyme in non-growing cells. (See text for explanation.)

(d) Methods for the study of enzyme synthesis, degradation, activation, and inactivation

At the present time, the methods which can be used to determine whether or not *de novo* synthesis of a particular enzyme is occurring fall into four categories:

1. the use of selective inhibitors,
2. immunology,
3. radioactive labelling, and
4. density labelling.

1. *Selective inhibitors.* Many micro-organisms, in particular fungi, produce substances which in low concentration kill competing organisms. These antibiotics have been

found to have a range of action, but most of them operate selectively, and specifically, on certain steps of RNA and protein synthesis. There is an enormous list of antibiotic inhibitors now available, but those that have been effectively used in plant systems are small in number. The commonest of these are shown in Fig. 8.6. Actinomycin D is a potent inhibitor of RNA synthesis in almost all organisms, but it is often necessary to use high concentrations because of permeability problems. Puromycin inhibits protein synthesis on the ribosome, but here again permeability problems exist with higher plant tissues. Cycloheximide is an extremely potent inhibitor of protein

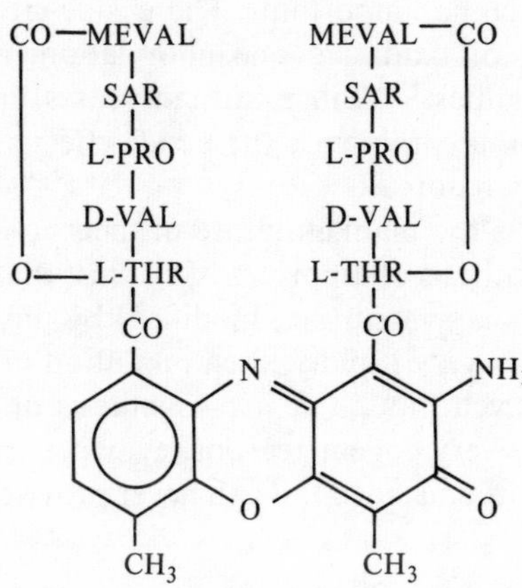

Actinomycin D

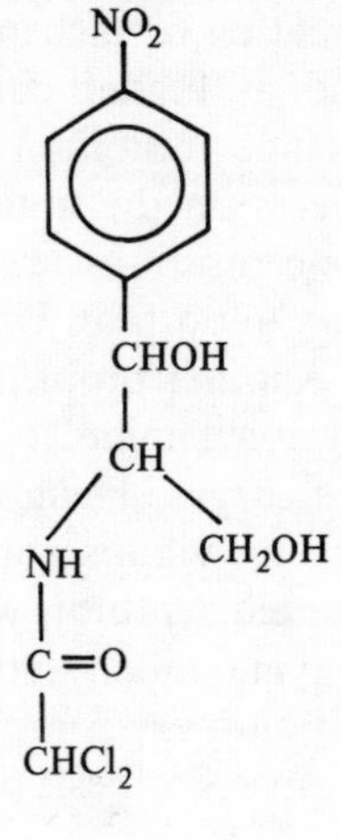

Chloramphenicol

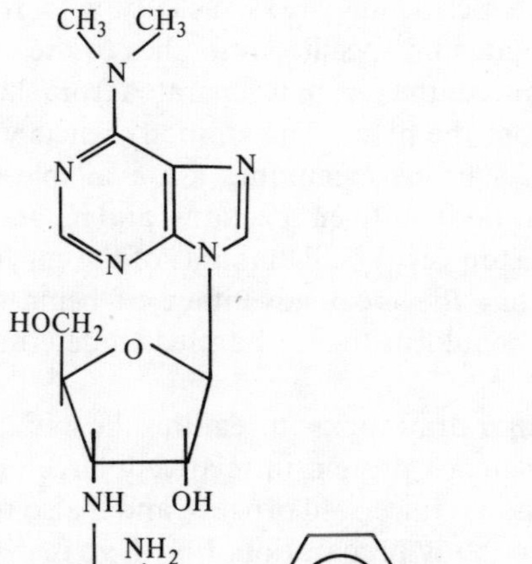

Puromycin

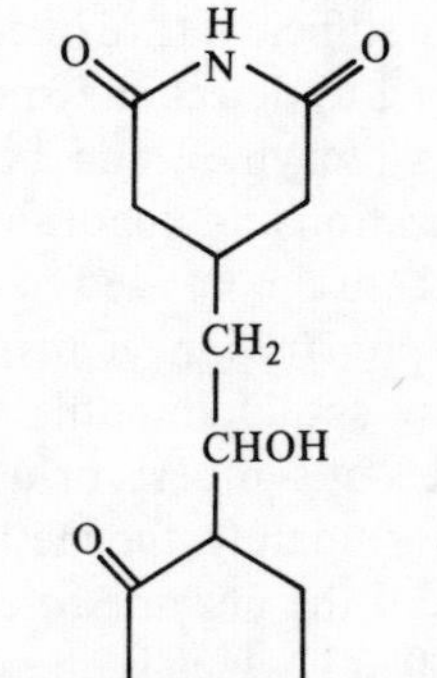

Cycloheximide

Fig. 8.6 The chemical structures of the four antibiotic inhibitors of nucleic acid and protein synthesis most popular with plant scientists.

synthesis and has the added advantage that it is highly specific for cytoplasmic ribosomes when used at low concentrations. Chloramphenicol, on the other hand, inhibits protein synthesis on chloroplast ribosomes.

The principle behind the use of inhibitors is that judicious applications of differentially selective substances should provide information on the locus of the light effect. For example, if cycloheximide inhibited a light-mediated increase in an enzyme whereas actinomycin D, and chloramphenicol did not, it could be reasonably postulated that the enzyme is synthesized on cytoplasmic ribosomes, and that the effect of light is on protein synthesis, rather than RNA synthesis. Unfortunately, the real situation is much more complex. None of the inhibitors is wholly specific in its action, and at higher concentrations side-effects become important. These side-effects include inhibition of respiration, uncoupling of oxidative phosphorylation, and effects on the uptake and transport of metabolites.[100] Since in many cases, high concentrations are necessary due to permeability problems, these side-effects are very important. Unfortunately, they are usually ignored.

Another problem with the use of inhibitors is the blanket nature of their action. Cycloheximide, for example, will inhibit the synthesis of all proteins, not just that of the enzyme being investigated. If an enzyme is activated (i.e., modified) by phytochrome via a separate enzyme, which turns over very rapidly, then inhibition of all protein synthesis will prevent the light effect, even though *de novo* synthesis of the enzyme under study does not occur. For these very cogent reasons, evidence from inhibitor experiments can only be regarded as circumstantial; it can never prove that *de novo* synthesis of an enzyme is occurring.

2. *Immunology*. The immune response of higher animals to foreign proteins in the bloodstream is the basis of probably the best method of determining whether or not synthesis of a specific enzyme is occurring. In this method, the enzyme is isolated and purified to a single homogeneous protein and is then injected into an animal's bloodstream, usually that of a rabbit. Over a period of weeks, usually with repeated injections of more enzyme, the animal builds up specific antibodies to the enzyme. When a sufficient titre of antibody is produced, the serum is separated from the blood and used to detect enzyme in extracts from the plant. The antibody reacts with the antigen (enzyme), and both precipitate. Various techniques are available for the visualization and quantification of the antibody/antigen reactions, and in favourable cases these can be used to detect and measure very small amounts of the enzyme in a crude preparation. Thus, this method has the great advantage of being able to measure actual amounts of enzyme, and could in theory be used to determine the net rates of synthesis or degradation.

Unfortunately, the method has two major drawbacks: it requires the initial purification of the enzyme to a single homogeneous protein in relatively large amounts (*ca*. 100 µg), which is an extremely tedious and uncertain process, and it also requires many weeks and sometimes months in antibody preparation. For these reasons, the method has only rarely been used with plant enzymes, and never in the investigation of a phytochrome-mediated response. It does, however, have more promise than any of the other methods to answer the important questions raised in this section, and its use is likely to increase in the near future.

3. *Radioactive labelling*. In this method, tissues are incubated in solutions of radioactive amino acids during the period when enzyme synthesis is suspected. The enzyme

is then extracted and purified again to a single homogeneous protein, whereupon measurement of the radioactivity in the purified enzyme is determined. If no other protein impurities are present, the radioactivity per unit protein is a measure of the synthesis of the enzymes which has taken place. Here again, the reliance upon complete purification makes the method extremely difficult and uncertain; furthermore, in comparative experiments it is necessary to achieve identical purifications in order to be able realistically to compare results.

In spite of these problems, the method has been used with some success in the photo-control of PAL in potato tubers, although phytochrome is apparently not the photo-receptor for this response (see below).

4. *Density labelling*. With the exception of inhibitor applications, this method is probably the simplest to carry out although here again the results are not always simple to interpret. The method is based on the assumption that if a protein is synthesized from amino acids, which are labelled with a stable heavy isotope (e.g., D or ^2H, ^{13}C, ^{15}N, or ^{18}O), then it should have a significant increase in mass without a concomitant increase in volume, i.e., its density will be higher.[208] If the enzyme is then centrifuged, at high speed, in a density gradient, it should finally equilibrate at a position in the gradient corresponding to its density, and thus it should be separable from the non-labelled, low-density enzyme. This method of isopycnic (i.e., equal density) equilibrium centrifugation using heavy salts, such as caesium chloride, for the gradient is capable of differentiating between proteins which differ in buoyant density by as little as 0·5 per cent.

One difficulty associated with this technique is the introduction of the density label into the amino acids. This is usually accomplished by incubating the tissues in D_2O, $H_2{}^{18}O$ or $^{15}NO_3^-$ under conditions in which either the hydrolysis of storage protein or the *de novo* synthesis of amino acids, is occurring. In this way deuterium, ^{18}O, or ^{15}N is incorporated into the amino acids, and thus into the proteins synthesized from them.

Density labelling was first used for plant enzymes by Filner and Varner[118] when they showed that α-amylase was synthesized, *de novo*, in barley half seeds treated with gibberellic acid. The method has also been used in studies of PAL responses to light, but as will be seen below, care must be taken in interpreting these results. Although density labelling is normally used to test for the existence, rather than to measure the rate, of *de novo* synthesis, the technique has been used to measure the rate of turnover of an enzyme (nitrate reductase[473]) and recent refinements in the technique suggest it might be possible also to measure rates of enzyme synthesis.[214]

From this account, it will be seen that no method yet available is able to demonstrate unequivocally that light, or any other factor, affects enzyme levels by directly controlling the rate of synthesis *per se*. The careful reader should bear this in mind when consulting the primary articles in this field, since the phrase 'enzyme synthesis' is often used quite casually to describe any increase in the extractable activity of an enzyme.

8.4   The photocontrol of lipoxygenase activity in mustard

Lipoxygenase catalyses the oxidation of unsaturated fatty acids containing a methylene-interrupted multiple-unsaturated system in which the double bonds are all *cis*.

Examples are linoleic and linolenic acids, and the action of the enzyme is to convert them to the conjugated *cis-trans* hydroperoxide.

In mustard cotyledons, lipoxygenase activity increases steadily during growth in the dark.[319] If continuous far-red light is given, there is no change in the time-course of increase in enzyme activity until just over 33 hours after the start of imbibition. At this point, far-red light completely, and suddenly, prevents all further increase in enzyme activity for a period of approximately 15 hours. At around 48 hours after imbibition, the enzyme activity begins to increase, even in far-red light, at a rate higher than that found in the dark (Fig. 8.7).

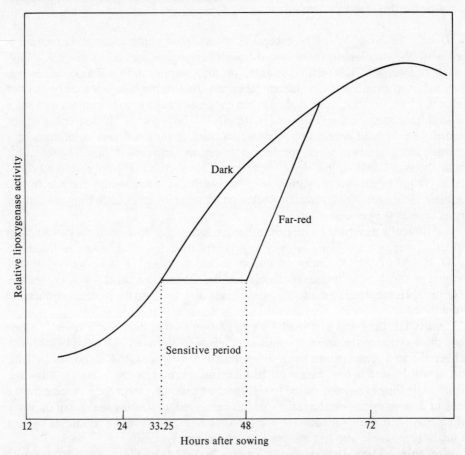

Fig. 8.7 The time-course of the changes in lipoxygenase activity in mustard cotyledons, during seedling growth in darkness or in continuous far-red light, from sowing. Note far-red light has no effect until 33·25 hours after which complete prevention of any further increase in activity is exerted for approximately 15 hours, when the activity increase resumes at a greater rate (after Oelze-Karow and Mohr).[319]

This phenomenon is clearly a most impressive case of the phytochrome regulation of enzyme activity. The response to far-red light is virtually instantaneous when given between 33·25 and 48 hours after the onset of imbibition; given at any other time, far-red light is totally ineffective. The mechanism which controls the level of lipoxygenase activity appears to go through a phase of sensitivity to phytochrome, and the transitions into and out of this phase appear to be very abrupt. It is also interesting that the enzyme activity of far-red treated plants 'catches-up' with that of the dark controls after the end of the sensitive period.

These considerations suggest that, during the time in which no increase in enzyme activity is observed, a late precursor, or perhaps an inactive form, of the enzyme, accumulates, thus allowing a more rapid attainment of the dark level after the end of the sensitive period. This is not, however, the interpretation favoured by the authors. Oelze-Karow and Mohr[319] conclude that phytochrome, in this case, is acting to repress the synthesis of lipoxygenase, and that this repression can only be exerted when other internal conditions of differentiation allow, i.e., during the sensitive phase. Repression implies action at the level of transcription, a very early step in enzyme synthesis, and one would not expect the cessation of transcription to be reflected in an immediate halt in enzyme synthesis, as is postulated. It seems safer to assume that phytochrome acts on a much later step in the processes causing the increase in enzyme activity allowing for the very rapid changes reported. These data and controversial conclusions are very important to an understanding of the control mechanisms in photomorphogenesis and would certainly repay further experimentation.

8.5 The photocontrol of phenylalanine ammonia-lyase (PAL) activity

Phenylalanine ammonia-lyase (PAL) catalyses the deamination of phenylalanine to cinnamic acid with the liberation of ammonia (Fig. 8.8). Phenylalanine ammonia-lyase has been found in a wide range of plants,[469] and has been extensively purified from potatoes[177] and maize,[281] where it has a molecular weight of around 320 000 daltons. It appears to have regulatory properties as evidenced by unusual kinetic behaviour, and by end-product inhibition.[9] It is an important enzyme in secondary metabolism since it catalyses the branching reaction from the main stream of protein synthesis leading to an enormous range of secondary products including cinnamic acid derivatives, simple phenols, lignin, acetophenones, coumarins, flavonoids, and *iso*flavonoids.

The extractable activity of PAL is subject to photocontrol in a wide variety of species (Table 8.2). The photoreceptor involved is often, but not always, phytochrome and in several cases the most marked increases in enzyme activity are due to the blue-absorbing photoreceptor. In non-etiolated tissues light appears to stimulate PAL levels through the photosynthetic pigments.

In etiolated tissues, the light-mediated rise in PAL activity tends to follow a similarly shaped time course, irrespective of the photoreceptor. In each case, the onset of irradiation is followed by a lag phase, after which a phase of steady increase in extractable PAL activity occurs. In many cases the peak of enzyme activity is followed by a rapid decline, often reaching the original level found in dark-grown seedlings.

Examples of this time course are given in Fig. 8.9. In these examples, the lag phases are of approximately equal duration (*ca.* 90 minutes), but the duration of the phases of increase and decrease is very different. In gherkin hypocotyls, maximum enzyme activity is achieved at *ca.* 4 hours with continuous blue light, followed by a rapid decline; in pea terminal buds, after a brief red light treatment, maximum activity is at 6–7 hours, again followed by a rapid decline, but not down to the dark level; and in mustard hypocotyls, with continuous far-red light, maximum activity is not reached for *ca.* 24 hours, and the subsequent decline is correspondingly much slower. Potato disks are an exception to this general pattern in that considerable activity increase occurs in the absence of light. This is probably due to a wound reaction.

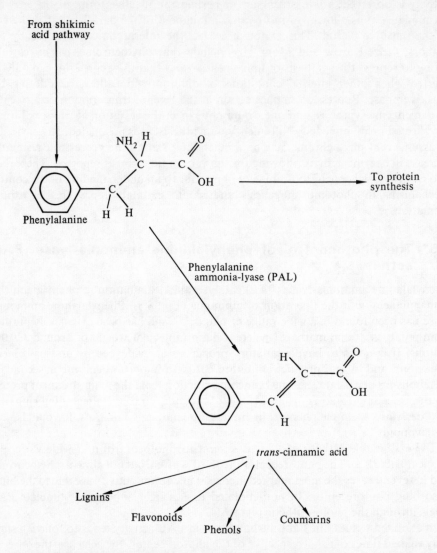

Fig. 8.8 The reaction catalysed by phenylalanine ammonia-lyase (PAL) showing the central position of the reaction in relation to the biosynthesis of a wide range of plant secondary products.

The general pattern of the response to light, irrespective of photoreceptor, is therefore as follows, and as shown digrammatically in Fig. 8.10:

1. a lag phase (usually *ca.* 90 minutes);
2. a phase of rapid increase in extractable activity; and
3. a phase of decline in extractable activity (not always seen).

Table 8.2 The photocontrol of phenylalanine ammonia-lyase and other enzymes of flavonoid biosynthesis

Region of pathway	Enzyme	Plant	Effect of light	Reference
Shikimic acid pathway	5-Dehydroquinase	Mung bean	No effect	2
	5-Dehydroquinase	Pea	Stimulation	2
	Shikimate: NADP-oxidoreductase	Mung bean	No effect	2
	Shikimate: NADP-oxidoreductase	Pea	Stimulation	2
	Shikimate: NADP-oxidoreductase	Pea	No effect	7
B-ring synthesis	Phenylalanine ammonia-lyase (PAL)	Potato	Stimulation	476
	PAL	Gherkin	Stimulation	101
	PAL	Buckwheat	Stimulation	377
	PAL	Mustard	Stimulation	97
	PAL	Pea	Stimulation	7
	PAL	Artichoke	Stimulation	317
	PAL	Cocklebur	Stimulation	478
	PAL	Strawberry	Stimulation	77
	PAL	Parsley	Stimulation	164
	PAL	Mung bean	Stimulation	2
	PAL	Red cabbage	Stimulation	105
	PAL	Radish	Stimulation	13
	Cinnamate-hydroxylase	Pea	Stimulation	364
	Cinnamate-hydroxylase	Buckwheat	Stimulation	3
	Cinnamate-hydroxylase	Soybean	Stimulation	163
	p-Coumarate: Coenzyme A-ligase	Parsley	Stimulation	161
	p-Coumarate: Coenzyme A-ligase	Soybean	Stimulation	163
A-ring synthesis	Acetate: Coenzyme A-ligase	Parsley	No effect	161
	Acetate: Coenzyme A-ligase	Soybean	No effect	162
Flavonoid modification	Chalcone-flavanone-isomerase	Parsley	Stimulation*	163
	UDP-apiose-synthetase	Parsley	Stimulation	164
Glycosylation	Apiosyl-transferase	Parsley	Stimulation*	163
	Glucosyl-transferase	Parsley	Stimulation*	163

* In these cases, the experiments were not designed to show whether light affected enzyme levels, but the results indicate that a stimulatory effect of light is likely.

In order to fully understand the response to light, it is clearly necessary to discover the processes going on in each of these three phases. Unfortunately, it must be said straight away that virtually nothing at all is known of those processes going on in the lag phase which, presumably, are essential for the increase in activity. Most effort has been applied towards an understanding of the phases of increase and decrease in enzyme activity.

(a) Phase of increase

Many workers have shown that inhibitors of protein synthesis will prevent the increase in PAL activity when applied during the lag phase or the early parts of the phase of increase.[3, 103, 356, 478] In a few cases, the inhibitor of RNA synthesis, actinomycin D, has been used, but only incomplete inhibition of the rise in PAL activity has been demonstrated.[98, 356a] Although these results are consistent with *de novo* synthesis of PAL during the increase phase, because of the problems of inhibitors referred to above, they by no means prove the case.

Attempts to demonstrate directly the *de novo* synthesis of PAL have been made by both the radiolabelling method, and the density labelling method. Zucker[478, 479] has reported the incorporation of radiolabel into PAL in *Xanthium* leaf disks. PAL

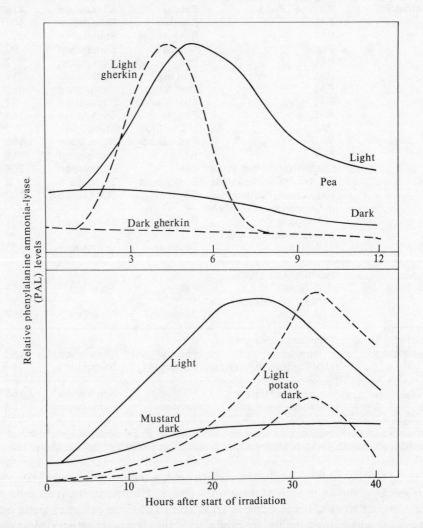

Fig. 8.9 The time-courses of the increases in extractable phenylalanine ammonia-lyase (PAL) activity following irradiation. Peas – red light (after Smith and Attridge)[403]; gherkins – blue light (after Engelsma)[101]; mustard – far-red light (after Rissland and Mohr)[356]; potato – white light (after Zucker).[475]

was purified to a single homogeneous protein to remove label associated with other proteins, but it was not specifically degraded to polypeptides to demonstrate the random incorporation of labelled amino acids. Thus, although this is highly suggestive evidence for PAL synthesis, it is not completely unequivocal. In addition, radiolabel was found in PAL from dark-treated disks, indicating that *de novo* synthesis was also proceeding in darkness. If this evidence is accepted as proof of synthesis, it means that light is not acting to *switch-on* synthesis, but merely to modulate either the rate of synthesis or the rate of degradation. Of these two possibilities, it seems most likely that light, acting in this case through the photosynthetic machinery, decreases the rate of PAL degradation, rather than increases its rate of synthesis, since Zucker found that the loss of radioactivity from preformed PAL was greater in disks incubated in darkness, than in disks incubated in the light.[479, 480]

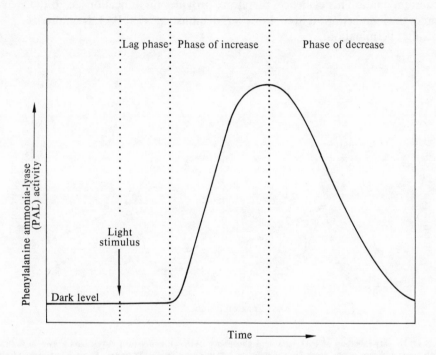

Fig. 8.10 Generalized *in vivo* response pattern of phenylalanine ammonia-lyase (PAL) to irradiation, showing the lag phase, the phase of increase, and the phase of decrease.

In a different tissue, potato tuber disks, PAL has been shown to be synthesized by density labelling. Sacher *et al.*,[367] rinsed disks in 97 per cent deuterium oxide, dried them, and repeated the treatment several times in order to ensure the uptake of significant amounts of deuterium oxide into the cells. They then incubated the disks in either white light or darkness. In both cases, the extracted enzyme had a significantly higher buoyant density than the PAL from water-incubated disks, showing unequivocally *de novo* synthesis of PAL in both light and dark-treated tissues. Here again therefore there is no question of light *switching-on* PAL synthesis and we must look either to a modulation of the synthesis and/or degradation rates, or to a mechanism of activation to account for the observed changes in extractable PAL activity.

Both of the above examples, *Xanthium* leaf disks, and potato tuber disks, are complicated since they almost certainly involve wound reactions – i.e., abnormal metabolic responses to the stresses imposed by slicing the tubers, or cutting the leaves. It may not be wise, therefore, to generalize on the basis of this information.

What is required is an analysis by density or radiolabelling of PAL in intact seedlings, which responds to light via either the blue-absorbing photoreceptor or phytochrome. In one published report[382] with intact mustard seedlings, the dry seeds were imbibed in deuterium oxide for over three days before the actinic light was given. At the end of the light period the enzyme was found to have a higher buoyant density than that from water-grown seedlings. It would indeed have been rather surprising if the enzyme had not been density labelled by this treatment, since only if it pre-existed in the dry seed could it have been synthesized in the absence of deuterium oxide. This evidence, therefore, provides no indications as to the mechanism of the light action; it does, however, demonstrate that PAL is synthesized some time after germination.

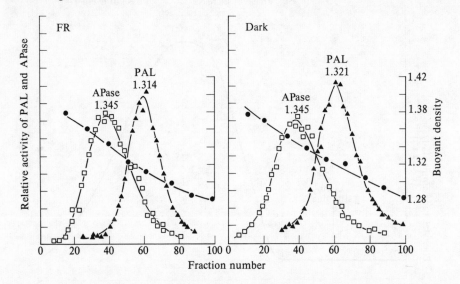

Fig. 8.11 Density-labelling of PAL and acid phosphate (APase) in mustard cotyledons grown in darkness for 48 hours and then either given far-red light, or left in darkness, for 48 hours. All seedlings were imbibed in H_2O and transferred to D_2O at 48 hours after sowing. The curves show the distribution of the enzymes in CsCl gradients after centrifugation to equilibrium. The buoyant densities of the enzymes are given above the peak positions.

N.B. The buoyant density of 'native' (i.e. non-deuterated PAL) is 1·295 kg/l and that of APase is 1·325 kg/l. Thus, although APase appears to be synthesized at the same rate in dark and far-red treated seedlings, PAL is apparently synthesized more rapidly in dark-grown seedlings. This result is explained in the text (after Attridge et al.).[6]

Other attempts have been made to carry out density-labelling tests in such a manner as to obviate the above criticisms. Attridge *et al.*,[6] working with intact mustard seedlings, imbibed seeds in water for 48 hours and then transferred them to 100 per cent deuterium oxide. The seedlings then were irradiated with far-red light for a further 48 hours and PAL extracted from the hypocotyls. Some of the various treatments and results are shown in Fig. 8.11. The striking feature to emerge from these experiments was that in all cases, the enzyme from the light-treated plants had a *lower* buoyant density than that from the dark-grown controls. Very similar results were also obtained with gherkin seedlings irradiated with blue light.

It is rather difficult to interpret these results in a manner consistent with a stimulation of *de novo* synthesis by light. The simplest conclusions seem to be that:

1. PAL is synthesized in darkness in both gherkin hypocotyls and mustard cotyledons;
2. PAL is also synthesized during blue light treatment of gherkin hypocotyls, and far-red light treatment of mustard cotyledons; and
3. light in both cases activates a pool of pre-existing inactive PAL molecules which were synthesized before deuterium was present in the tissues, thus lowering the mean buoyant density of the enzyme extracted from the light-treated plants.

Density labelling, therefore, appears to provide good evidence for an effect of light, acting in one case through phytochrome and in the other via the blue-absorbing photoreceptor, on the activation of pre-existing inactive PAL, and there is further

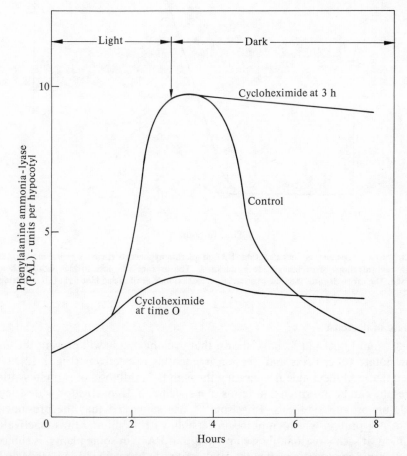

Fig. 8.12 The effect of cycloheximide given at the onset of irradiation, or at the time of maximum phenyl-alanine ammonia-lyase (PAL) activity, in gherkin hypocotyls. Note the antibiotic appears to prevent both the increase, and the subsequent decrease, in enzyme level (after Engelsma).[102]

evidence for this view, as will be seen below. PAL is synthesized during the light treatment, however, and there is no way of knowing as yet whether light *also* stimulates the rate of enzyme synthesis in addition to causing enzyme activation.

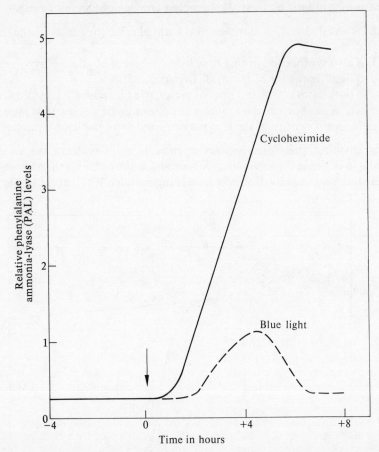

Fig. 8.13 The enormous increase in extractable PAL of gherkin hypocotyls elicited by spraying the seedlings with high concentrations of cycloheximide in darkness. The average response to blue light is shown for comparison. The arrow denotes the time of spraying cycloheximide or of giving blue light (after Attridge and Smith).[8]

(b) Phase of decrease

Zucker[477] and Engelsma[102] have shown that cycloheximide will prevent the loss of PAL in potato tuber disks and gherkin hypocotyls respectively (Fig. 8.12). If it is assumed that cycloheximide is operating through the inhibition of protein synthesis, these results can be interpreted in terms of a specific PAL-inactivator whose activity depends on protein synthesis. Engelsma[104] has suggested that the presumed inactivator is a protein which combines reversibly with PAL, and thus inactivates it. In addition, it seems reasonably well proven that PAL, in some plants, is subject to relatively rapid inactivation after the peak of the light-mediated increase has been reached. In gherkin, a subsequent incubation of the tissues at 4°C for 24 hours, followed by a return to 25°C in darkness, leads to a second rise in extractable PAL activity which cannot be prevented by high concentrations of cycloheximide.[104] This second rise, therefore, is likely to be due to activation of existing, inactive enzyme molecules. Thus, under certain conditions, PAL can exist in gherkin tissues in an inactive form.

Recently, it has been shown that a considerable pool of inactive PAL is present in the hypocotyls of gherkin seedlings grown in complete darkness from imbibition.[8] The evidence for this view is based on the fact that high concentrations of cycloheximide elicit an enormous rise in extractable PAL in dark-grown seedlings (Fig. 8.13). The increase in PAL due to cycloheximide is very much greater than that due to light treatment suggesting that the pool of inactive PAL is sufficiently large to provide all the molecules which become detectable with light treatment.

These results show that it is quite feasible that light may act by activating all, or some, of the PAL already present in the tissues in an inactive form. The activation process might require the synthesis of a specific enzyme, and in this case protein synthesis inhibitors would prevent the light-mediated increases in PAL. In another tissue, etiolated peas, slight differences in the properties of PAL extracted from dark-grown and light-treated seedlings have been found supporting the concept of light-activation.[9]

8.6   Nucleic acid and protein synthesis

(a) Nucleic acid synthesis

The gene expression hypothesis of phytochrome action has been a strong force in determining the approach of many investigators interested in delving into the biochemistry of photomorphogenesis. It is not surprising, therefore, that some considerable effort has been expended (and largely wasted) in searching for light-mediated changes in nucleic acid and/or protein synthesis occurring with a very short lag phase. The gene expression hypothesis, as set out by Mohr (see page 108), however, implies that the changes in gene transcription will be relatively small, and that stimulation of the transcription of one set of genes might occur in association with inhibition of the transcription of other sets of genes. Thus, if phytochrome were to act in the way proposed in the hypothesis, only very small changes in the rate of synthesis of total messenger-RNA would be expected. Messenger-RNA comprises only about five per cent of total cell RNA, and at present there are no reliable techniques for separating messenger-RNA from the rest. The separation of individual messenger-RNAs from each other is an even more intractable problem. In addition, it is now known that ribosomal-RNA and transfer-RNA (the other two major RNA fractions) are synthesized in growing cells at rates at least as high as is messenger-RNA, and thus it is not possible to distinguish messenger-RNA synthesis on the basis of its rapidity, as was earlier thought.

With hindsight, therefore, it can be said that attempts to obtain evidence relating to the initial mode of action of phytochrome, by investigating the effects of light on nucleic acid synthesis, were doomed from the start. Early investigations were restricted to the measurement of total nucleic acid levels in seedlings given actinic light treatment. For example, in mustard cotyledons treated with continuous far-red light, a considerable increase in total RNA was observed over 24 hours.[453] This increase had a lag phase of 6 hours, and thus could only be considered to be a secondary effect of the primary changes. In any case, it is likely that the bulk of the increase was due to increased numbers of ribosomes rather than to changes in messenger-RNA synthesis.

Later work has concentrated on shorter term experiments designed to test for rapid effects on the synthesis of specific fractions of cellular nucleic acids. As mentioned above, however, the fractionation methods do not, as yet, allow separation and quantitization of the most interesting RNAs. Dittes and Mohr[87] pursued the mustard cotyledon effect by incubating tissues in radioactive precursors of the nucleic acids for 20 minutes after the onset of far-red light. The nucleic acids were then extracted and separated by chromatography on columns of methylated albumin on kieselguhr (MAK). No differences could be observed in any fraction of nucleic acid from the columns. This result may indeed reflect the true situation, but, as the authors pointed out, MAK chromatography is a very insensitive technique which gives only poor separation of the major types of nucleic acids.

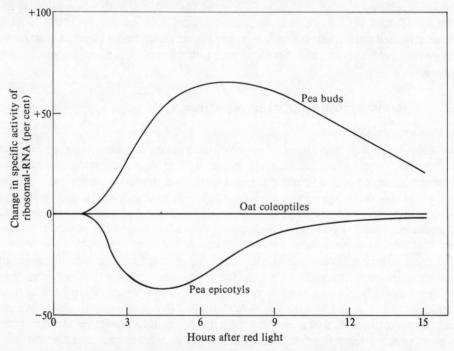

Fig. 8.14 The effect of 5 min red light on the incorporation of ^{32}P into the ribosomal-RNA of pea epicotyl sections, intact pea buds and apical sections of oat coleoptiles (after Koller and Smith).[250]

In another investigation, Koller and Smith[250] measured the rates of synthesis of ribosomal-RNA in three photomorphogenic test tissues and concluded that no causal relationship existed between the effects on nucleic acid synthesis and the subsequent developmental changes. RNA was separated here by acrylamide gel electrophoresis which enables complete separation of all ribosomal-RNA species but not of messenger-RNAs.[267] In the terminal buds of etiolated peas, where brief red light leads to a large increase in leaf growth, ribosomal-RNA synthesis was significantly stimulated, starting about 2–3 hours after light treatment and peaking at about 6 hours (Fig. 8.14). In stem sections of etiolated peas, on the other hand, where red light inhibits extension growth, a corresponding decrease in ribosomal-

RNA synthesis occurred. Finally, in apical sections of oat coleoptiles, where red light stimulates extension growth, but does not increase cell number, no effects on ribosomal-RNA synthesis were detectable. These results suggest that the observed changes in ribosomal-RNA were associated with the subsequent growth changes and merely reflected the synthesis of new ribosomes as new cells were formed. It was shown that 5-fluorouracil, a metabolic analogue of uracil, which completely inhibits ribosomal-RNA synthesis, did not prevent the red light induced growth changes in any of the three test materials. It was thus concluded that the phytochrome effects on ribosomal-RNA synthesis were not essential prerequisites for the ultimate developmental changes.

Phytochrome has also recently been shown to cause an increase in the proportion of ribosomes present as polysomes. In etiolated bean leaves, brief red light causes a three-fold increase in the proportion of cytoplasmic polysomes, starting about 2 hours after light treatment and reaching a new steady state at about 10 hours.[329] At the same time, a greater increase in plastid polysomes occurs. In both cases, the effects were reversible by brief far-red light, proving the involvement of phytochrome. These results, which were obtained by computer analysis of the complicated curves obtained from sucrose density-gradient centrifugation of ribosome preparations, confirm the earlier findings of Williams and Novelli[460, 461] who showed that white light led after a 2 hour lag to a marked rise in the polysomal proportions of ribosomes from etiolated corn leaves.

Pine and Klein,[329] in considering the mechanism of the increase in polysomal proportions, rejected the possibilities that light causes either a sudden assembly of ribosomes onto pre-made messenger-RNA strands, or a piling-up of extra ribosomes onto pre-existing polysomes. The first possibility would imply a very short time-lag, and the second an increase in the average number of ribosomes per polysome, neither of which were found. They concluded that the most likely mechanism was that the messenger-RNA became more stable, i.e., its synthesis continued at the same rate while its degradation was slowed. This would allow for a higher proportion of polysomes on a steady-state basis, but the length of time required to accumulate a new stable level of polysomes might be several hours. Although this conclusion cannot be definite until direct measurements of messenger-RNA stability can be made (and such measurements are at present impractical), it nevertheless could fit in quite well with other completely unrelated facts of photomorphogenesis which are described later in this chapter.

Other workers have shown red/far-red light effects on enzymes of RNA metabolism. Bottomley[42] isolated nuclear and plastid RNA-polymerase from etiolated pea seedling terminal buds previously treated with light. Both red and far-red light increased the plastid polymerase activity by about 50 per cent over the dark control, and far-red reversibility of the red effect was not detected. Both red and far-red light also increased the activity of the nuclear polymerase, but red light was more effective, and subsequent far-red light reversed the effect down to the far-red control value. Thus, the nuclear polymerase activity appears to be increased via a phytochrome-mediated mechanism, whereas although the plastid polymerase activity is also increased by light, phytochrome is apparently not the sole photoreceptor. The increases in polymerase activity did not become detectable until about 4 hours after light treatment, and it was therefore concluded that they were not manifestations of the primary action of phytochrome.

Increases in the activity of ribonuclease, an enzyme which degrades RNA, also occur in response to phytochrome photoactivation, and with a similar time course. Acton[1] has shown, using lupin seedlings, that a ribonuclease isozyme, associated with the ribosomes, is significantly increased by red light after a lag of about 4 hours. The effect is red/far-red reversible. Thus, both enzymes of RNA synthesis and RNA degradation are controlled by phytochrome, but probably only as secondary responses.

(b) Protein synthesis

Little direct investigation of the possible phytochrome control of protein synthesis has been carried out. This is rather surprising in view of the manifold effects of phytochrome on enzyme activities. However, there is considerable evidence on the effects of white light on protein synthesis, and some tantalizing indications of the mechanism of the effect. Unfortunately, most of this work has been carried out by biochemists interested in the control of protein synthesis and unconcerned about the nature of the photoreceptor. The white light treatments have commonly been relatively short (*ca.* 2–3 hours) and the effects occur within this time period. They are therefore quite rapid compared with many of the known biochemical manifestations of photomorphogenesis. The plants, however, are usually subjected to high irradiances during this period and the possible involvement of photosynthesis, or at least of light perception by the photosynthetic photoreceptors, looms large. Whether these effects can therefore be attributed to phytochrome, to the blue-absorbing photoreceptor, or to protochlorophyll/chlorophyll is as yet anybody's guess. Nevertheless they are of sufficient interest and importance for them to be outlined here.

Effects of white light on total protein levels were first observed in 1954 long before phytochrome was discovered.[84] Ten years later Williams and Novelli[460] showed that ribosomes prepared from white light treated maize, bean and soybean leaves had a greatly enhanced capacity to incorporate ^{14}C-leucine into a protein-like product *in vitro*. The lag phase in maize shoots was about 2 hours and maximum effect was achieved by 3 hours. The light effect appeared to be an intrinsic property of the ribosomes since the same supernatant fraction (i.e., containing transfer-RNAs, aminoacyl transfer-RNA-synthetases, etc.) was used in all cases.

Subsequently, the same authors investigated this response further by assaying the capacity of ribosomes to support the incorporation of phenylalanine into polyphenylalanine using polyuridylic acid (poly(U)) as an artificial messenger.[461] Again, light did not affect the capacity of the soluble supernatant to support phenylalanine incorporation, but did increase the ribosomal activity by about 35 per cent. Thus, the effect was again shown to be due to an intrinsic property of the ribosomes. Recently, this work has been followed up by Travis *et al.*[439] who showed that 3 hours of white light enhanced the ability of maize shoot ribosomes to incorporate phenylalanine into polypeptide, using poly(U) as messenger, by about 100 per cent depending on the magnesium ion concentration. Light-grown plants yielded ribosomes with a much lower magnesium ion concentration optimum than dark-grown plants.

In order to determine the nature of this difference in ribosomal properties, the effects of puromycin were tested. Puromycin, an antibiotic inhibitor of protein synthesis (see Fig. 8.5) attaches to ribosomes and releases a peptidyl-transfer-RNA complex that is essential for the binding of phenylalanine-transfer-RNA to ribosomes at low magnesium ion concentrations.[69] When the light-treated ribosomes were incubated in puromycin, the high rates of phenylalanine incorporation at low magnesium ion concentration were lost, and the ribosomes appeared identical to those from dark-grown plants, although puromycin depressed the activity of the 'dark' ribosomes. Thus, it appears that one of the effects of light is to enhance the amounts of a ribosome-associated peptidyl-transfer-RNA which enables phenylalanine incorporation to occur at higher rates. This peptidyl-transfer-RNA is one of a range of substances necessary for the commencement of protein synthesis on the ribosome, and thus is known as an initiation factor.

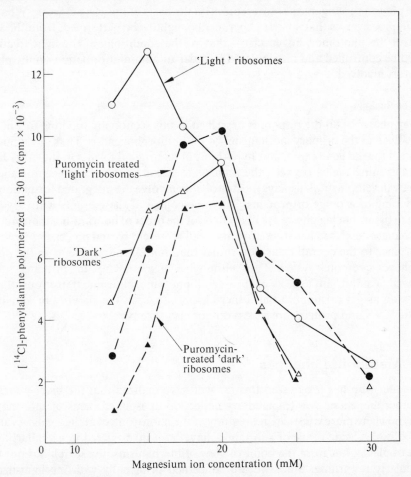

Fig. 8.15 Protein synthesizing capacity of ribosomes from dark-grown and 3-hour white light-treated maize leaves. Note that the optimal magnesium ion concentration is decreased by the light treatment. Puromycin treatment of the ribosomes from light-treated plants appears to convert them to 'dark-ribosomes' as far as magnesium ion concentration is concerned (from Travis et al.).[439]

Other initiation factors may also be involved in the light response. When ribosomes were 'washed' in high concentrations of potassium chloride, which removes associated proteins, but not the peptidyl-transfer-RNA, the 'light' ribosomes showed a much higher magnesium ion concentration optimum and were reduced in activity, but were still considerably more active than the 'dark' ribosomes. When ribosomes were both potassium chloride-washed and puromycin-treated, both magnesium ion concentration optima and activities were identical. These results are summarized in Fig. 8.15.

The conclusion to be derived from these findings is that light may activate protein synthesis by direct effects on the ribosomal machinery, particularly on one or more initiation factors which are more or less loosely bound to the ribosomes. In view of the considerations mentioned above concerning the photoreceptors for this response, it would be extremely dangerous to draw conclusions about the mechanism of action of the photomorphogenic photoreceptors on the basis of this elegant work. What can be said, however, is that similar experiments ought to be performed, from the view-point of the photomorphogenecist – that is, the conditions of the light treatment should be controlled and manipulated in order to obtain information on the photoreceptors involved.

(c) Conclusions

The lag phases of all the responses described in this section are too long for them to be involved in the primary mechanisms of photomorphogenesis. Thus, no support is provided for the gene expression hypothesis in its original form. On the other hand, the intervening events between the immediate action of the photoreceptor and the ultimate developmental displays most probably involve the integrated formation and functioning of a range of proteins. The time-course of these events will vary from tissue to tissue, but lag phases are likely to be of the order of hours rather than minutes. The changes described here therefore may well represent partial processes of essential importance to the overall transition rather than initial, causal, events in the photomorphogenic response system. Thus, although changes in RNA and protein synthesis may well be important aspects of the overall photomorphogenic transitions, there is no reason as yet to suppose that other changes (e.g., in the stability or activity of specific RNA and protein molecules) do not also take part.

8.7   Metabolite changes

The reader who has progressed this far will have realized that the biochemistry of photomorphogenesis is a fragmentary collection of isolated items of information. This is nowhere more true than in the study of the photocontrol of the levels of various metabolites. A wide range of substances have been investigated in a similarly wide range of plants, but from the point of view of mechanisms, the search has not been particularly rewarding. Since we are concerned principally with mechanisms, no attempt will be made to give encyclopaedic coverage of this confused subject. There are, however, a few cases in which attempts have been made to obtain evidence relating to mechanisms. Those substances studied in most detail in this respect have been flavonoids, ascorbic acid, and carotenoids.

(a) Flavonoids

Flavonoids are secondary products whose basic structure is shown in Fig. 8.16. The molecule is made up of two aromatic rings, the A-ring being synthesized by head-to-tail condensation of acetate units and the B-ring from the shikimic acid pathway via phenylalanine (see page 178). There are several different classes of flavonoids distinguished by the oxidation level of the bridge carbons; Fig. 8.16 shows the basic flavan nucleus (1), the flavonol structure (2) and the anthocyanidin structure (3). The flavonols and anthocyanidins are the two most important classes of flavonoids, and are found in an extremely wide range of plants in many of which their synthesis is controlled by light (Table 8.3). Flavonols and anthocyanidins are commonly present as glycosides, the glycosides of anthocyanidins being known as anthocyanins. Related substances also often controlled by light are the phenylpropanoids, which are C_6–C_3 compounds biosynthetically equivalent to the B-ring plus the bridge carbons of the flavonoids.

The photocontrol of flavonoid levels is exemplified by the situation in mustard seedlings where, after a lag of about 3–4 hours, continuous far-red light causes a massive increase in the level of anthocyanin (Fig. 8.17).[292] In this figure, the effects

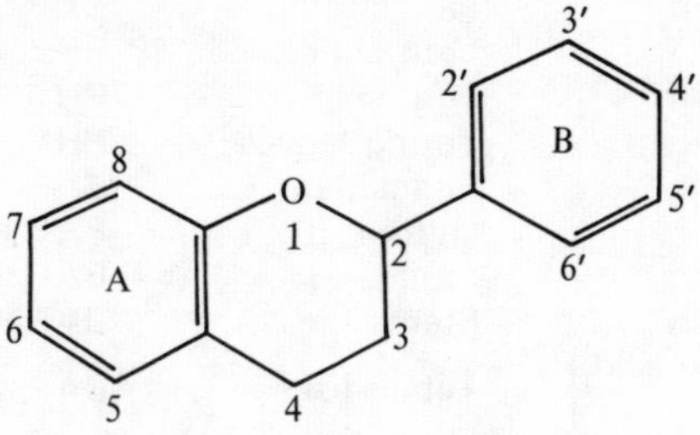

(1)

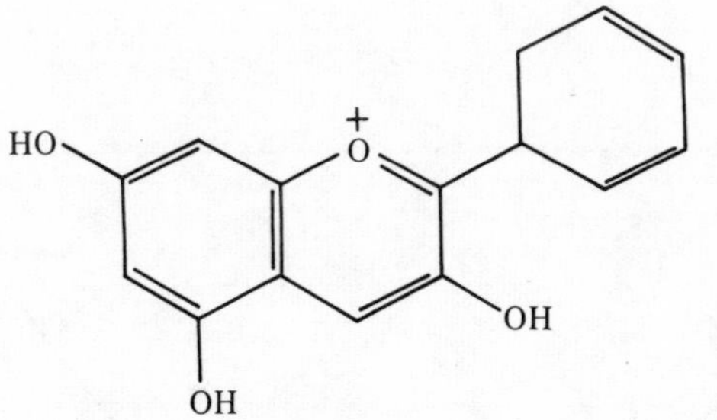

(2)

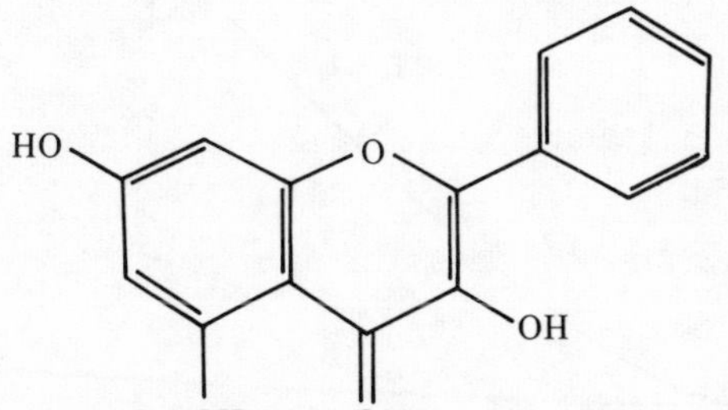

(3)

Fig. 8.16 The basic structure of flavonoids:
 1 – the flavan nucleus
 2 – the anthocyanidin structure
 3 – the flavonol structure

Table 8.3 The operation of the basic photoresponse systems in the regulation of flavonoid and phenylpropanoid levels.
(PHY = phytochrome; BAP = blue-absorbing photoreceptor.)

Plant	Substances	Organs	Photoreceptor	Reference
Mustard	Anthocyanin	Hypocotyl	PHY	292
		Cotyledon	BAP	
Gherkin	Phenylpropanoid	Hypocotyl	PHY	106
	Flavonol	Cotyledon	BAP	
Turnip	Anthocyanin	Hypocotyl } Cotyledon }	PHY, BAP	388
Pea	Flavonol	Young leaves	PHY, BAP	404
Buckwheat	Flavonol } Anthocyanin }	Shoots	PHY, BAP	440
Sorghum	Anthocyanin	First internode	BAP	92
Tomato	Flavonol	Fruit cuticle	PHY	330
Apple	Anthocyanin	Fruit skin	PHY, BAP	389
Bean	Anthocyanin	Young leaves	PHY, BAP	465
Maize	Anthocyanin	Endosperm culture	BAP PHY (?)	413
Spirodela	Anthocyanin	Fronds	BAP, PHY (?)	425
Impatiens	Anthocyanin	Stems	BAP, PHY	316
Potato	Chlorogenic acid } Flavonol }	Tuber discs	BAP, PHY	475
Red cabbage	Anthocyanin	Leaves	BAP, PHY	99

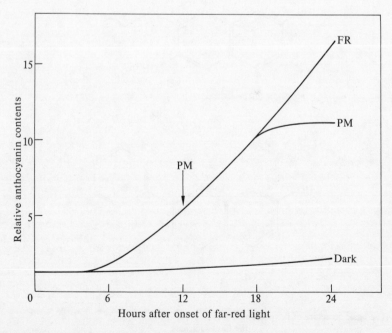

Fig. 8.17 The effect of puromycin (PM) on the formation of anthocyanins in mustard seedlings in continuous far-red light. The arrow indicates the time of application of the inhibitor (after Mohr and Senf).[306]

of puromycin (an inhibitor of protein synthesis) can be seen. Puromycin (Fig. 8.6) inhibits synthesis in the presence of continuous far-red light irrespective of the time at which the inhibitor is applied. The effects of the inhibitor take several hours to manifest themselves, and this has been taken to indicate that continued anthocyanin synthesis is dependent on one or more critical enzymes whose half-life is of the order of a few hours. The obvious candidate for the critical enzyme is PAL, but although PAL activity is regulated by light (as described in detail on page 182), the levels of PAL are not always closely correlated with the rates of anthocyanin synthesis. It thus appears that other enzymes in the pathway may also be limiting the rate of anthocyanin synthesis under certain conditions.

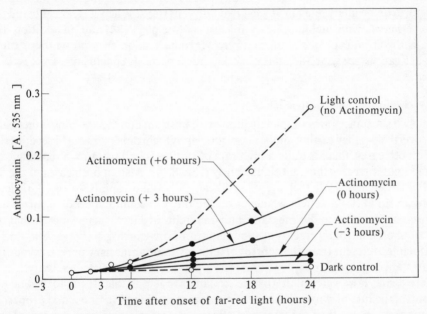

Fig. 8.18 The effect of actinomycin D given at various times before, or after, the onset of far-red light, on the accumulation of anthocyanin in the mustard seedling (from Mohr and Bienger).[300]

Mohr and colleagues[300] have also investigated the effects of actinomycin D (see Fig. 8.6) on phytochrome-mediated anthocyanin synthesis in some detail, and have concluded that specific induction of messenger-RNA synthesis is involved. The essential data upon which this view is based is shown in Fig. 8.18. A pre-incubation of the seedlings in actinomycin D, for 3 hours before the onset of the irradiation, completely abolishes the light-mediated synthesis of anthocyanin. However, a 3 hour incubation in actinomycin D (in darkness) given after the start of irradiation, only partially inhibits the increase in anthocyanin. From these and related experiments, it was concluded that actinomycin D could block the primary processes of the light response if given at or before exposure of the tissues to the light stimulus, but permits continued synthesis of anthocyanin if the synthetic process had begun before application of the inhibitor. This is generally consistent with the concept that light acts to selectively switch-on the transcription of the genes coding for the limiting enzymes in anthocyanin synthesis. There are, however, other possible interpretations.

The requirement for pre-incubation in actinomycin D in order to fully inhibit the light effect has led Smith[400] to suggest that the inhibitor acts by preventing synthesis of a messenger-RNA that is necessary for light action. On this view, actinomycin D would need to be added sufficiently long before the light treatment for the pre-existing messenger-RNA to be removed by degradation. The function of phytochrome would thus be to affect either the messenger-RNA or the protein synthesizing machinery, in some way, so that a greater rate of enzyme synthesis could be achieved. In this respect, this interpretation of the data fits in with what is known of the changes in ribosome properties brought about by continuous white light, and described in the previous section.

The effects of protein and nucleic acid synthesis inhibitors on anthocyanin synthesis therefore can be interpreted in at least two different ways. When it is remembered that all experiments with such inhibitors are suspect (see page 174) and, in addition, that there is as yet no evidence at all that phytochrome acts on enzyme synthesis rather than activation (see page 183), it is clear that no respectable conclusion can be reached as yet.

(b) Ascorbic acid and carotenoids

These two substances are completely unrelated, both chemically and biosynthetically, but nevertheless they exhibit similar control by phytochrome in mustard seedlings.[20,] [383] In both cases, there is quite a high rate of synthesis in dark-grown seedlings, but far-red light increases this rate significantly (Fig. 8.19). Also in both cases, if the far-red light is switched off the rate of synthesis falls quite rapidly to the dark level. Actinomycin D has no effect whatever on the rate of accumulation of ascorbic acid or carotenoids either in darkness or in far-red light. Mohr[297] has proposed that this type of photocontrol represents a 'modulation' of pre-existing activities, and thus is, in principle, different from the control of anthocyanin synthesis, where in favourable circumstances phytochrome 'switches-on' a previously non-existent activity.

Here again, however, an alternative explanation is possible. If phytochrome acts to control the rate of synthesis of specific enzymes by changing ribosomal properties, as suggested above, then the observed effects on ascorbic acid and carotenoid synthesis could easily be accounted for. It would be necessary to assume that the messenger-RNAs for the as yet unknown limiting enzymes in the two pathways were quite stable over long periods; if this were the case, actinomycin D would have no effect on the rates of final product formation. On this view, it is no longer necessary to distinguish between modulatory and differentiational effects of phytochrome.

8.8   The photocontrol of chlorophyll synthesis

The plastid is *the* characteristic plant organelle; no other organisms contain plastids. The term plastid covers a wide range of structurally different, but basically inter-convertible, bodies of varying function. The most important, both numerically and metabolically, is the chloroplast, which contains the photosynthetic machinery of the cell. Other plastids include the starch-storing amyloplasts, the flower-colour produc-ing chromoplasts and the etioplasts of the dark-grown plant. Plastids are normally about 5 μm in diameter, although their overall shape appears to be relatively flexible and, in some cases, mobile. It is known that plastids contain their own discrete types of DNA, ribosomes and transfer-RNA,[239] and that at least some of the

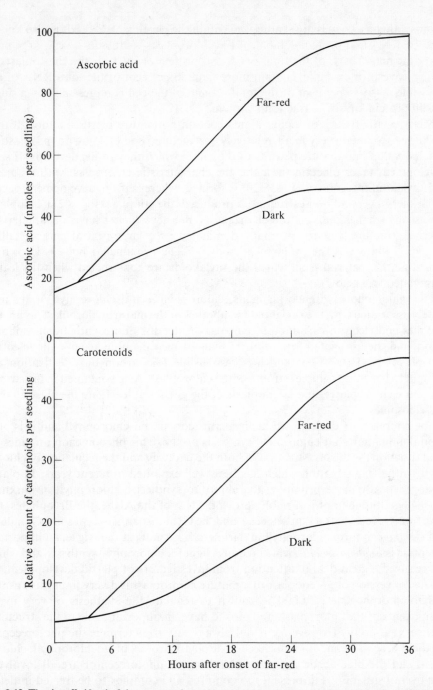

Fig. 8.19 The virtually identical time-course of ascorbic acid (top) and carotenoid (bottom) synthesis in mustard seedlings grown in darkness or continuous far-red light (after Bienger and Schopfer[20]; and Schnarrenberger and Mohr).[383]

photosynthetic enzymes are synthesized within the plastids.[44, 45, 396] It is also known that the RNA and protein content of the chloroplasts represents a very large part, if not the major part, of the total RNA and protein of the cell; the chloroplasts of barley leaves, for example, contain more than 50 per cent of the total RNA of the cell while up to 35 per cent of the total protein of the cell is represented by a single plastid protein known as fraction I protein.[398]

Meristematic tissues of higher plants contain no mature plastids, although they are known to contain very small, relatively unstructured bodies, known as proplastids, which develop into mature plastids as the cell in which they reside develops. If such development takes place in the light, the characteristic chloroplast with its grana and thylakoid membranes (Fig. 8.20) is formed, whereas if the development occurs in darkness a very different structure is produced, the etioplast (Fig. 8.21). Etioplasts are usually smaller than mature chloroplasts, but nevertheless have a characteristic internal structure which is normally dominated by a large central paracrystalline formation known as the prolamellar body.[159] The prolamellar body is known to constitute the material from which the thylakoids are constructed when the leaf is illuminated (see below).

The major photosynthetic pigments, chlorophyll *a* and chlorophyll *b*, are intimately associated with the thylakoid membranes of the mature chloroplast, as are the enzymes and electron-carrying proteins responsible for photosynthetic phosphorylation, and the associated formation of reduced nicotinamide adenine dinucleotide phosphate (NADPH). The enzymes responsible for carbon dioxide fixation are thought to be in the 'soluble' stroma of the chloroplast. Also contained in the stroma of mature chloroplasts, and of etioplasts of the grasses at least, are the plastid DNA and ribosomes.

The etioplasts of dark-grown angiosperms contain no chlorophyll and it is only upon illumination that chlorophyll synthesis occurs. This phenomenon provides an ideal situation for the investigation of both the pathway and the regulation of chlorophyll synthesis, a situation which has been well exploited in recent years. In plants phylogenetically more primitive, the ability to synthesize chlorophyll in darkness is usually, although not invariably, present. Most of the Algae, the Bryophytes, the Pteridophytes except the Equisetaceae, and the Gymnosperms except some Cycadales and *Gingko* can form chlorophyll in darkness.[239] Amongst the Algae, an important exception is *Euglena gracilis*, which requires light for chlorophyll synthesis, and which has received a great deal of attention from investigators of plastid development.[239]

The conversion of an etioplast to a mature chloroplast is a very important aspect of photomorphogenesis. Light is known to regulate the synthesis of very many components of the chloroplast, and also to have an important role in the structural changes that occur. However, it is not always easy to determine the photoreceptor for these processes, and it seems likely that phytochrome, protochlorophyll, chlorophyll, and the blue-absorbing photomorphogenic photoreceptor are all involved. The overall situation is at present too confused and complex to be treated in detail here and we will be concerned with principles only.

(a) Chlorophyll synthesis

The pathway of biosynthesis of chlorophyll *a* and chlorophyll *b* is summarized in Fig. 8.22. For most of its length the pathway is common to that of the synthesis of the haem group, which has been very fully investigated in animals in connection with

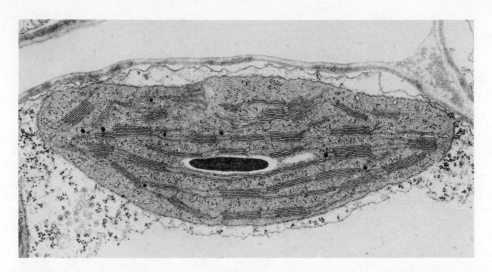

Fig. 8.20 Electron micrograph of a mature chloroplast of a wheat leaf (print kindly supplied by Dr B. E. S. Gunning).

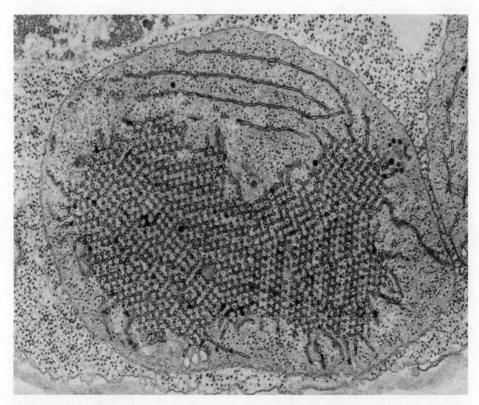

Fig. 8.21 Electron micrograph of an etioplast in a wheat leaf (print kindly supplied by Dr B. E. S. Gunning from ref. 419a).

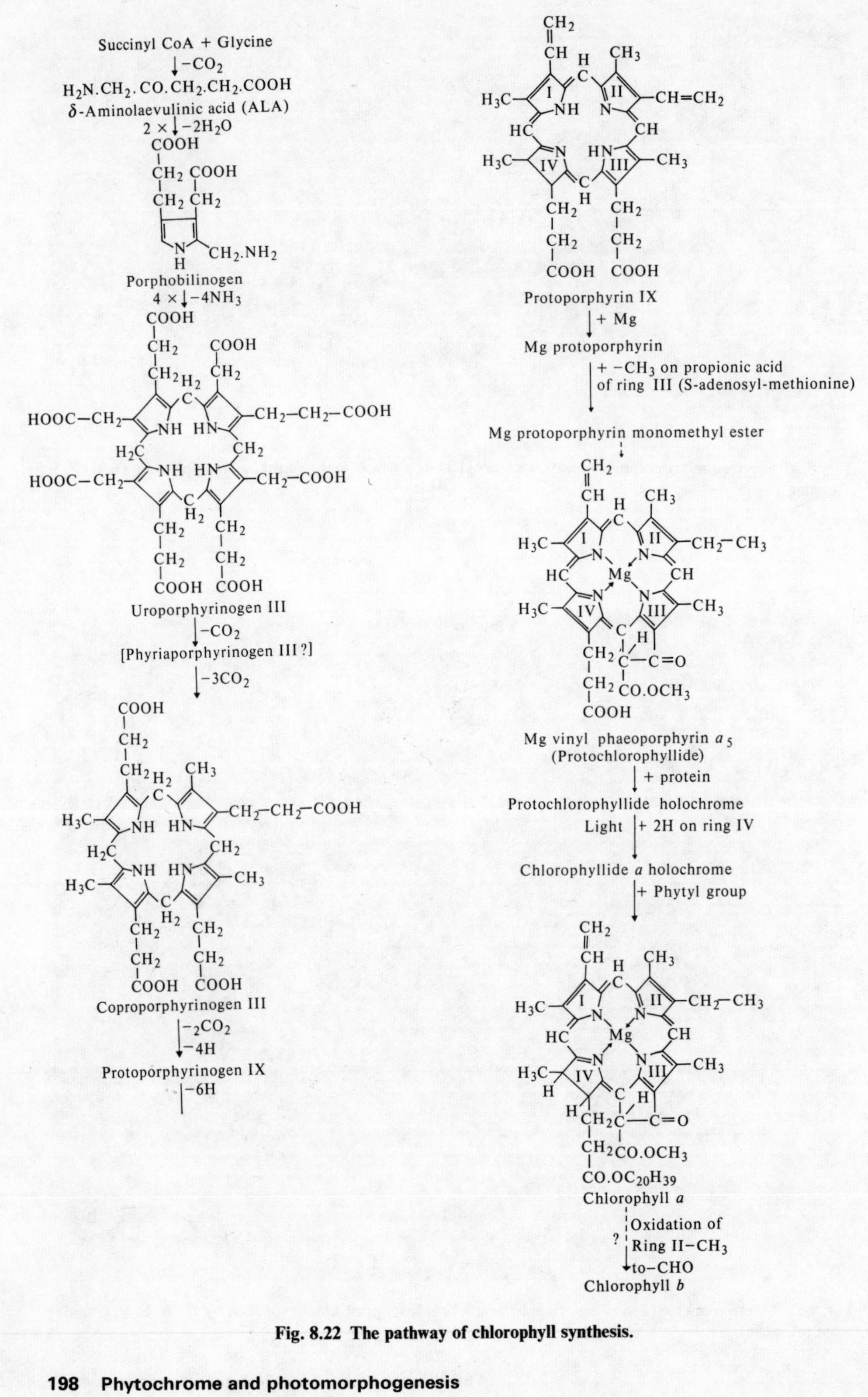

Succinyl CoA + Glycine
↓ −CO$_2$
H$_2$N.CH$_2$.CO.CH$_2$.CH$_2$.COOH
δ-Aminolaevulinic acid (ALA)
2 × ↓ −2H$_2$O

Porphobilinogen
4 × ↓ −4NH$_3$

Uroporphyrinogen III
↓ −CO$_2$
[Phyriaporphyrinogen III?]
↓ −3CO$_2$

Coproporphyrinogen III
↓ −2CO$_2$
↓ −4H
Protoporphyrinogen IX
↓ −6H

Protoporphyrin IX
↓ + Mg
Mg protoporphyrin
↓ + −CH$_3$ on propionic acid
of ring III (S-adenosyl-methionine)

Mg protoporphyrin monomethyl ester

Mg vinyl phaeoporphyrin a_5
(Protochlorophyllide)
↓ + protein
Protochlorophyllide holochrome
Light ↓ + 2H on ring IV

Chlorophyllide a holochrome
↓ + Phytyl group

Chlorophyll a
? ↓ Oxidation of Ring II−CH$_3$ to −CHO
Chlorophyll b

Fig. 8.22 The pathway of chlorophyll synthesis.

haemoglobin synthesis. As far as chlorophyll is concerned, the pathway has been almost fully worked out with the exception of some essential details which, as it happens, relate particularly to those steps that appear to be regulated by light.

The first reaction in the pathway is one of those that are subject to doubt since it has not been possible as yet to isolate from leaves the enzyme that catalyses the reaction. An enzyme is prevalent in animals and microorganisms which catalyses the synthesis of δ-aminolaevulinic acid (ALA) from succinyl-Coenzyme A and glycine, and which is therefore called ALA-synthetase. All attempts to isolate this crucial enzyme from leaves or chloroplasts have so far met with failure,[237] except for one, as yet, unconfirmed report, presented in an abstract only, of ALA-synthetase activity being present in spinach leaf extracts.[289]

Evidence for the view that ALA-synthetase is the enzyme responsible for ALA synthesis in wheat leaves comes from the finding that radioactivity from ^{14}C-succinate and ^{14}C-glycine is incorporated into the porphyrin ring of the chlorophyll, albeit at a very low specific activity.[357] Another possible route for ALA synthesis is by the transamination of γ, δ-dioxovaleric acid, using, probably, glutamine, alanine, or phenylalanine as preferential amino group donors:

$$CHO.CO.CH_2.CH_2COOH + R.CH(NH_2).COOH \xrightarrow[\text{(ALA-transaminase)}]{}$$

γ,δ-dioxovalerate $\qquad\qquad\qquad$ α-amino acid

$$\longrightarrow H_2N.CH_2.CO.CH_2.CH_2COOH + R.CO.COOH$$

δ-aminolaevulinic acid $\qquad\qquad$ α-oxoacid.

This enzyme has been found in *Chlorella* by Gassman *et al.*[143]; its participation in chlorophyll synthesis, however, has not been proved nor has it yet been found in higher plants. It will become clear later that the nature of the mechanism for ALA synthesis is of crucial importance for studies into the photomorphogenic control of chlorophyll formation.

The remaining steps in the pathway up to the final conversion of protochlorophyllide to chlorophyll are reasonably well worked out (see ref. 237) and do not appear to be subject to photocontrol, and therefore need not concern us. The photoconversion of protochlorophyllide to chlorophyll, as presently understood, is however a highly complex process involving the association of the chromophore with a specific protein, known as the holochrome, the esterification of the propionic acid residue of ring IV with the long chain alcohol phytol to form protochlorophyll, and also the reduction of ring IV in a photochemical reaction.

This photochemical reaction is a crucial point in chlorophyll synthesis since, in the absence of light, this reaction cannot proceed, and thus the synthesis is blocked at the level of protochlorophyll or protochlorophyllide. However, it is not yet clear in which order the three reactions outlined above occur, or in fact whether the order is necessarily the same in all cases. In the review by Kirk[237] it is pointed out that at least two, and probably three, spectrally distinct forms of protochlorophyllide are known to exist, and that the sequence of events in the photochemical conversion of these components to chlorophyll could involve as many as four spectrally distinct forms of chlorophyllide (chlorophyllide is chlorophyll without the phytyl residue). The light dependent steps as proposed by Kirk occur in the sequence shown in Fig.

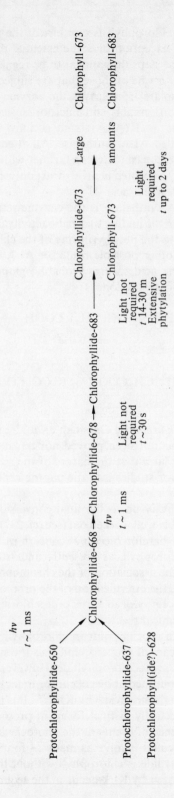

Fig. 8.23 Partial processes in the photochemical conversion of protochlorophyllide to chlorophyll (from Kirk[237] with permission from 'Biochemical Aspects of Chloroplast Development'. Annual Review of Plant Physiology, Vol. 21, p. 17. Copyright © 1970 by Annual Reviews Inc. All rights reserved.

8.23. The numbers after the chemical names refer to the absorption peaks exhibited by the substances. It is suggested that the differences in absorption spectra shown by two variants of (presumably) the same substance, are due to different degrees of association of the molecules in the thylakoid membranes. Implicit in all this is the view that the formation of the thylakoid membranes must be associated with the final photochemical steps in chlorophyll synthesis. Thus, although in the strict sense, the photochemical conversion of protochlorophyllide to chlorophyll is not a photomorphogenic reaction, it nevertheless appears to be a necessary event for the developmental process of membrane formation in the chloroplast.

The action spectrum for the photoconversion of protochlorophyllide (or protochlorophyll) to chlorophyll is very similar to the absorption spectrum of protochlorophyll (the presence of the phytyl residue has little effect on the visible spectrum) and thus it seems clear that protochlorophyll(ide) is the *in vivo* photoreceptor for this reaction. The photoreduction of protochlorophyllide also requires the action of the holochrome, probably in a catalytic manner. Protochlorophyllide is attached to the holochrome by non-covalent bonds, and it is not yet clear whether or not the holochrome serves as the proton source for the photoreduction reaction.

(b) Photoconversion of etioplast to chloroplast

The materials of the prolamellar body (Fig. 8.21) make up the thylakoid and grana membranes of the mature chloroplast. This structural reorganization occurs in response to light and can be divided into three phases. The first step is the transformation of the regular crystalline lattice of the prolomellar body into a disorganized structure in which the tubules, although still connected to each other, are not regularly oriented.[159] Tube transformation (as this process is known) and protochlorophyll photoconversion have similar action spectra and energy requirements.[243, 449] They are both very rapid reactions and can be accomplished by a single flash of 1 ms duration.[221] These and other lines of evidence suggest that the photoreceptor for this first photoreaction in chloroplast formation is protochlorophyll itself.[14, 185] If, after a flash of actinic light, leaves are returned to darkness, the prolamellar body gradually recrystallizes.[14] It seems likely that this recrystallization may be associated with the regeneration of protochlorophyll (see below).

The next stage of chloroplast development is the extrusion of the tubular material into the stroma region forming two-dimensional double membranes which appear to be perforated in many places. This process also requires light and appears to be irradiance dependent.[220] With high irradiances the whole process can occur within a few minutes. Normally, however, it takes 2–3 hours. The action spectrum shows a marked peak at 450 nm, with very little activity at all in the red and far-red regions,[184] indicating that the photoreceptor cannot be either protochlorophyll, chlorophyll, or phytochrome. As the irradiation treatment continues, the pores in the membranes disappear, usually forming complete sheets within about 2–3 hours of light treatment.[159] During this time, no membrane material has been manufactured *de novo*, all of it having been derived from the prolamellar body.

The final stage of ultrastructural conversion is correlated with the major phase of chlorophyll synthesis. This consists of the *de novo* duplication of the membrane material to form the grana as seen in mature chloroplasts (Fig. 8.20). This process also needs light and is apparently irradiance dependent, although the nature of the photoreceptor is unknown. The duplication process obviously cannot begin until tube extrusion has occurred, and thus it is subject to a lag phase of 2–3 hours after the start of irradiation. This is similar to the lag phase in chlorophyll synthesis discussed below, and as with chlorophyll synthesis, the duration of the lag can be significantly reduced by pretreating the leaves with red light some hours before exposing them to white light. If this is done, the rate at which grana appear is substantially enhanced.[14] This phenomenon is known to be, at least partially, controlled by phytochrome, as is described below.

Thus, the conversion of the etioplast to the chloroplast is a highly complex sequence of events in which protochlorophyll, phytochrome and an unknown blue-absorbing photoreceptor all appear to play a role. The early stages of tube transformation and extrusion do not appear to be greatly dependent on metabolic energy, but the later stages of grana accumulation are probably considerably energy dependent. It thus seems likely that photosynthesis itself will have an increasingly important role as the overall process reaches completion.

(c) The lag phase and the control of chlorophyll synthesis

When seedlings are grown in darkness it would be expected that protochlorophyllide (or protochlorophyll) would accumulate in the leaves. In fact the *in vivo* concentration of protochlorophyll(ide) in etiolated leaves, as measured by direct spectrophotometry of the living leaves, remains quite steady at a very low level. When etiolated barley leaves are irradiated with a very brief flash of intense light, in order to photoreduce their protochlorophyllide, protochlorophyllide resynthesis resumes after a lag of about 5–10 minutes and continues until the level of protochlorophyllide reaches the original dark level, whereupon synthesis stops.[10] It has been suggested by Granick[152] that protochlorophyllide regulates its own synthesis by feed-back inhibition of ALA-synthetase. This suggestion has also been made by other workers[238] and is supported by Granick's own finding that excised etiolated shoots of barley when fed with exogenous ALA for 1–2 days accumulate protochlorophyllide.[151]

When etiolated leaves are placed in continuous white light, although there is an instantaneous conversion of protochlorophyllide to form a small amount of chlorophyll, there is normally a considerable lag phase before the major phase of chlorophyll synthesis occurs. This lag can be of the order of 2–4 hours in certain cases. In 1956, Withrow et al.,[467] using etiolated bean leaves, showed that the lag in chlorophyll formation could be virtually abolished by pretreatment of the leaves with a very small amount of red light given between 5 and 15 hours *before* the transfer to continuous white light. It was also found that the effect of the red light pretreatment could be reversed by immediate subsequent irradiation with far-red light (Fig. 8.24). Virgin in 1961 produced an accurate action spectrum for the elimination of the lag phase in wheat leaves showing a prominent maximum at 660 nm with shoulders on the lower wavelength side of this maximum. The evidence thus suggests that the effect of pre-irradiation with red light on the elimination of the lag phase in chlorophyll synthesis is a classical phytochrome phenomenon. This conclusion should, perhaps, be approached with a little caution, since protochlorophyllide also absorbs strongly at 660 nm and is probably photoreduced even by the very low energy pre-irradiation treatments. Furthermore, complete reimposition of the lag phase by far-red light is very difficult to achieve and in most cases only 40–60 per cent reversal is obtained (Fig. 8.24), although Mohr[297] reports complete far-red reversal in mustard. The general lack of reversibility may be due to one or more of several factors, such as

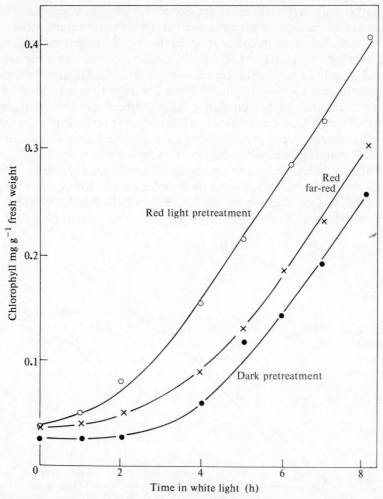

Fig. 8.24 Reduction of the lag phase in chlorophyll synthesis in wheat leaves by prior irradiation with brief red light, and its partial reversion by far-red light (unpublished results).

rapid escape from phytochrome control, extreme sensitivity to low concentrations of Pfr, or some effect dependent directly on protochlorophyllide absorption in addition to phytochrome.

It has been suggested that phytochrome acts, over a long period, to elevate the levels of ALA-synthetase (or ALA-transaminase, whichever is responsible for the formation of ALA), thus removing the block on chlorophyll synthesis at the beginning of the pathway. Evidence for this comes from two sources. In 1958, Virgin[447] found that etiolated wheat leaves which had been pretreated with red light several hours before they were given a brief flash of intense white light to photoreduce their protochlorophyllide, resumed protochlorophyllide synthesis immediately, rather than after a 5–10 minute lag as observed in fully dark-grown seedlings. If the hypothesis of protochlorophyllide acting as a feedback inhibitor of ALA synthesis is accepted, then a shortening of the lag phase in the resumption of protochlorophyllide synthesis would most likely be due to an increased level of the enzyme responsible for ALA synthesis, rather than to changes in the activity of that enzyme. Supporting evidence for this view comes from the observation that the red light pretreatment effect on the elimination of the lag phase in chlorophyll synthesis can be replaced by supplying etiolated leaves with exogenous ALA. Upon transfer to continuous white light, leaf segments fed in such a way show no lag phase in greening, and there is no detectable effect of a pretreatment with red light.[395] These findings are in agreement with the view that phytochrome acts by stimulating ALA synthesis, possibly by causing an increase in the level of ALA-synthetase (or ALA-transaminase), but it is also possible that phytochrome photoactivation leads to an increased supply of precursors of ALA synthesis, namely succinyl-Co-enzyme A and glycine.

This problem was intensively investigated in etiolated barley seedlings by Nadler and Granick[312] in 1970. When 9-day-old etiolated leaves were fed with glycine and/or succinate, there was no decrease in the lag phase of chlorophyll synthesis upon subsequent exposure to white light, suggesting that the intermediates for ALA

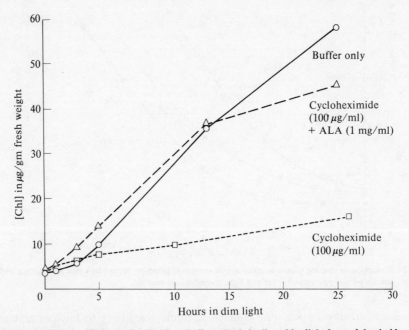

Fig. 8.25 The inhibition of light-mediated chlorophyll synthesis in dim white light by cycloheximide, and its reversal by δ-aminolaevulinic acid (ALA) (from Nadler and Granick).[312]

synthesis were not limiting. If, however, similar seedlings were fed with ALA and transferred to *dim* white light (to prevent photobleaching) the lag phase was eliminated. Furthermore, chlorophyll synthesis in dim light continued in the presence of ALA even when inhibitors of protein synthesis such as cycloheximide were also applied; similar doses of cycloheximide in the absence of ALA feeding prevented chlorophyll synthesis (Fig. 8.25). These data suggest that 9-day-old etiolated leaves contain all the enzymes responsible for converting ALA to chlorophyll and that their levels are non-limiting. Also, the conversion of exogenously fed ALA to chlorophyll during the first few hours of light treatment shows that the holochrome protein, which photoreduces protochlorophyllide to chlorophyllide, is also non-limiting.

Both cycloheximide, and chloramphenicol prevent chlorophyll synthesis in the absence of exogenous ALA. Inhibition of chlorophyll synthesis in bean leaves by chloramphenicol has also been observed.[483] Chloramphenicol given at the time of red light pretreatment led to a reimposition of the lag phase in greening, while application at the onset of the continuous white light, although depressing the rate of synthesis, did not reimpose the lag phase. A similar result was observed by Nadler and Granick[312] with cycloheximide.

Nadler and Granick[312] have also applied various inhibitors of nucleic acid synthesis in attempts to determine whether light acts through inducing RNA synthesis. No inhibition of chlorophyll synthesis was observed with 5-fluorouracil (which is relatively specific for ribosomal-RNA synthesis), with mitomycin-C, or with various analogues of the purine and pyrimidine bases. A certain amount of inhibition was observed with actinomycin D, which also inhibits messenger-RNA synthesis, but only when the tissues were preincubated with the antibiotic for 24 hours before illumination.

These experiments of course, are not immune to the criticisms applied here to other work based on inhibitors of protein and nucleic acid synthesis. They are however, quite different in one sense since the most important conclusions are derived from situations in which application of the inhibitors did not inhibit the synthesis of chlorophyll; thus actinomycin D only inhibited after a long pretreatment and cycloheximide had no effect in the presence of ALA. Thus, it can be said that nucleic acid synthesis is probably not required for the light effect on chlorophyll synthesis, and that ALA allows chlorophyll synthesis in the absence of protein synthesis. These conclusions are valid even if the antibiotics are inhibiting other processes *in addition* to nucleic acid and protein synthesis.

From these results, Nadler and Granick[312] have proposed a model scheme for the control of chlorophyll synthesis in barley (Fig. 8.26). The control on the rate of ALA synthesis, and thus of chlorophyll, is thought to depend, at least in part, on the synthesis of a limiting protein that forms ALA, and on the rate of its degradation. This protein is presumably ALA-synthetase. The role of phytochrome on this model is to activate stored messenger-RNA to make the requisite proteins for ALA synthesis. This proposal leads directly from the failure of the applied nucleic acid synthesis inhibitors to prevent chlorophyll synthesis. The synthesis of the proteins apparently requires the participation of both cytoplasmic and plastid ribosomes since chlorophyll synthesis is prevented, and the lag phase reinstated, by both cycloheximide and chloramphenicol. Other evidence not described here indicates that the ALA synthesis system is relatively unstable and that breakdown proceeds with a half-life of approximately $1\frac{1}{2}$ hours. After formation of ALA, the production of chlorophyll proceeds unchecked by the action of the remaining enzymes of the biosynthetic chain, these enzymes being non-limiting and stable for at least 6 hours. The final steps in synthesis are dependent on light via the photochemical reduction of protochlorophyllide by the holochrome.

8.9 Control mechanisms in photomorphogenesis

The aim of this chapter has been to present some, at least, of the information relating to the biochemical changes which occur in cells undergoing a photomorphogenic transition, in the hope that one or more unifying trends might emerge. To attempt

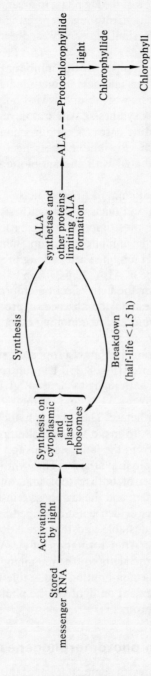

Fig. 8.26 The hypothetical scheme proposed by Nadler and Granick[312] for the control of chlorophyll synthesis in barley.

to make hard and fast conclusions on the basis of the available evidence would, indeed, be foolhardy, but it does appear as if some negative conclusions are possible and that some predictions might logically be made.

First of all, it must be concluded that all biochemical changes caused by light, other than those occurring within the first seconds or minutes of light treatment, must be secondary effects of the primary reaction of the photoreceptor. This is particularly obvious for phytochrome, but is probably also true for the blue-absorbing photoreceptor. Thus, we have a situation in which the absorption of light by the photoreceptor triggers an enormous range of metabolic responses, each proceeding with different time-courses.

When the individual secondary processes are considered in detail, we are almost always forced to the conclusion that the overall rate of biosynthesis of a certain metabolite is determined by the amount of activity of limiting enzymes in the pathway. The really important questions therefore relate to whether the changes in the activities of these limiting enzymes are due to increased synthesis, decreased turnover, or to modifications of pre-existing enzyme molecules.

At the present time, it can be quite categorically stated that light has not been shown unequivocally to increase the rate of *de novo* synthesis of a single enzyme. This is due mainly to the lack of satisfactory techniques for measuring the rate of synthesis of single enzymes. For one enzyme, PAL, there is evidence that light activates pre-existing inactive enzyme molecules, but there is no reason why this mechanism should be a general mechanism of light action. Indeed, it seems most likely from the wealth of circumstantial evidence, from inhibitor studies mainly, that the synthesis of many enzymes may be stimulated by light. If it is, we are faced with the problem of deciding whether the control is exerted at the level of transcription, or translation. The results of the actinomycin D experiments on anthocyanin, ascorbic acid, carotenoid, and chlorophyll synthesis described above all have one thing in common: complete inhibition cannot be obtained when actinomycin D is given at the time of light treatment, and the antibiotic must be applied some hours before to achieve complete inhibition. Thus, there is no good evidence to support an effect of light on transcription. On the other hand, much work points to a change in the properties of either the messenger-RNA or the ribosomes, such that enhanced synthesis of the critical enzymes can occur.

If this is the real situation, and at present it is only speculation supported by a few unrelated facts, it means that photomorphogenic transitions are basically quantitative changes in the expression of sections of the genetic information already being expressed, rather than the qualitative switching-on and switching-off of new sections of the genome. This concept is thus in direct contrast to the gene expression hypothesis as originally put forward by Mohr in 1966. It would, however, be premature to advance either the concept of translational regulation or that of enzyme modification as an alternative hypothesis.

Thus, we are left in the familiar situation of confusion. Future progress will almost certainly be dependent on refinements in the techniques for measuring the rates of synthesis of specific enzymes. The discovery of a method for estimating the rates of synthesis of specific messenger-RNAs would obviously be of major importance, but such a method is unlikely to be found in the near future.

9
Conclusions and speculations

9.1 The mechanisms of photomorphogenesis

Photomorphogenesis is a paradigm of development in multicellular organisms. All the elements of developmental control are displayed, ranging from the perception of an environmental stimulus, through its transduction via rapid biophysical and biochemical changes within the perceptive cells, to the ultimate alteration of the patterns of morphogenesis. The very size of this 'introductory text' is a reflection of the enormous amount of information on photomorphogenesis that has been assembled in the last 20–30 years. It therefore behoves us to ask two questions:

1. do we now have a preliminary understanding of the mechanisms of photomorphogenesis; and
2. does our knowledge of photomorphogenesis help us in coming to an understanding of the general nature of the mechanisms underlying development.

As for the first question, the answer seems to be a rather hesitant and querulous 'yes', at least if we think in principles instead of details. Light is perceived by at least two photoreceptive substances; phytochrome, which has been isolated and comprehensively characterized, and a blue-light absorbing pigment whose chemical nature can, as yet, only be speculated upon. The primary effects of the absorption of actinic radiation by phytochrome may be related to membrane properties, and the ultimate developmental effects may be results of the initial alteration in the compartmentation of metabolites caused by the change in membrane properties. In these later events, we see changes in the *in vivo* levels of specific enzymes, although it cannot yet be decided how the observed changes in enzyme levels are brought about.

The concentration of our ideas on membranes for the immediate action of phytochrome bears close analogy with the other major processes in photobiology, i.e., photosynthesis and vision. In both of these cases, the active photoreceptors appear to be components of membranes which are oriented with respect to the direction of the incident light. In this way, the photoreceptors themselves are also oriented, thus making maximum use of the available radiant energy. We do not know, as yet, whether phytochrome is an actual component of plant cell membranes, or whether it is unattached yet acts upon them in some way, nor do we know which membranes are involved in photomorphogenesis. We must assume, however, that regulation of membrane properties is an integral step in the photomorphogenic reaction sequence.

If higher plants are similar to *Mougeotia*, phytochrome is likely to be a component of the plasmalemma and to change its axis of orientation upon photoconversion. If this is proved to be true, then photomorphogenesis and vision indeed bear a remarkably close analogy. The vision photoreceptor, rhodopsin, is known to exist in several different forms in which the chromophore exists as different isomers; in one form the protein opsin and the chromophore, in fact, fall apart. Some of the possible forms of rhodopsin may be reversibly photoconverted by different wavelengths of light in exactly the same way as phytochrome. The rhodopsin molecules are arranged in disk-like plates in the rod cells of the retina with their chromophores oriented parallel to the disk membranes and perpendicular to the long axis of the rod. When viewed along the rod, however, they appear to be randomly oriented.[58] Upon photoactivation, the molecules of rhodopsin apparently 'flip-over' in a way that is consistent with their acting as a transport factor across the membrane in which they are situated.[74] This is a very close analogy to the transport factor hypothesis of phytochrome action described in chapter 5.

Although we do not yet have any real understanding of the primary changes in membrane properties, nevertheless the biggest gap in our knowledge of phytochrome action is the time lag that occurs between the pigment photoconversion/membrane changes, and the earliest biochemical changes that can be associated with the later developmental events, e.g., the increase in enzyme activities. This lag is in most cases at least 60 minutes and, at present, virtually nothing is known of the processes occurring in that time period. There are perhaps some indications that changes in ATP/ADP concentrations and/or gibberellin levels may occur within this first hour, but it is not yet known how general these responses are, nor how they can be related to the later developmental processes. Rather more information is available on the photoregulation of enzyme levels, but since these changes are definitely secondary effects of phytochrome action, we are unlikely to learn a great deal about the primary mechanism of photomorphogenesis by such studies.

It is precisely these later events in photomorphogenesis, however, which are likely to tell us something about the more general problem of the regulation of plant development. It is almost an axiom nowadays that organismal development involves the selective control of gene expression, the real question being the mechanism through which the control is exerted. It has often been expected, on purely *a priori* grounds, that gene expression will be found to be regulated principally at the transcription step. As pointed out in chapter 8, there is as yet not a single item of evidence that unequivocally proves transcriptional control of gene expression during photomorphogenesis. We can go even further, and state that most of the acceptable evidence supports post-transcriptional control. There are, in fact, some cogent reasons for believing that, at least in certain cases, control may be exerted after translation, on the activation of previously synthesized inactive enzymes. At first sight, such a mechanism might appear to be rather expensive in energetic terms, but looked at from the point of view of the plant it may represent a very valuable property that enables it to adapt rapidly to changing environmental conditions. The synthesis of new enzymes in higher plant cells is probably rather a slow process when compared to that in bacteria – in fact no real information is available on the rates of true *de novo* enzyme synthesis in the cells of intact plants. The existence of pools of inactive forms of critically important enzymes which can be rapidly activated in response to environmental stimuli could well be of immense adaptive value, and could therefore have been selected for during evolution, in spite of the energetic penalties involved. In any case, it seems extremely unwise at the moment, to state categorically that photomorphogenic control of gene expression is exerted solely at one level or another – it may well be that different control mechanisms operate for different genes depending on the requirements for rapid or slow responses.

As this final chapter is speculative, the author may perhaps be allowed the indulgence of putting forward his own favourite idea as to how phytochrome works. This view involves the operation of a 'second messenger' by which the primary action of phytochrome within certain important membranes is linked to the various later effects on enzyme formation and activation. The concept of a second messenger in phytochrome action was first put forward by Jaffe in relation to the role of acetylcholine (see ref. 212 for review), and this seems the only way to reconcile all the available information. The evidence supporting a general role for acetylcholine as a second messenger, however, does not appear to be very strong (see page 163). Nevertheless, the concept of a second messenger finds a parallel with the action of cyclic-AMP in animal cells.[217] The scheme, as far as phytochrome is concerned, is shown diagrammatically in Fig. 9.1, in which the unknown second messenger (X) is considered to be liberated from a storage compartment in accordance with the hypothesis described on page 118 (Fig. 5.20); after release it is capable of reacting with several different reaction partners so that gene expression is regulated at translation and enzyme modification, and possibly also at transcription. This hypothesis, while admittedly speculative, has two advantages: it can account in general terms for all the functions of phytochrome; and it has definite heuristic value, and should

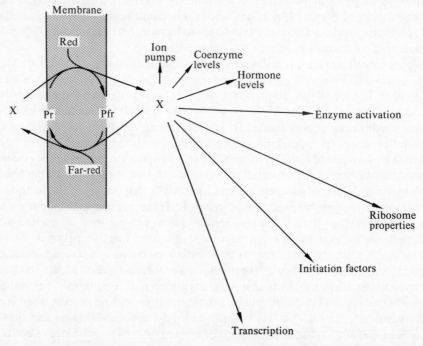

Fig. 9.1 A speculative model for the action of phytochrome in regulating rapid metabolic processes, and longer term developmental process.

Active phytochrome is considered to be located in one or more critical membranes, in which it acts to control the transport of a second messenger, X, from one side to the other; X is then able to interact with many processes, the time lags being roughly denoted by the lengths of the arrows. Compare with Figs. 1.2 and 5.20.

direct attention towards the further intensive investigation of the interaction of phytochrome with membranes, and the possibility that phytochrome might have binding sites for small metabolites.

9.2 Possibilities for future research

The development of the phytochrome story, from the first brilliantly deductive experiments of H. A. Borthwick and S. B. Hendricks, has been conditioned by the invention of new techniques. The early Beltsville work, so dependent upon the elaboration of highly accurate action spectra, would probably not have been possible without the construction of the enormous spectrograph with which the visible spectrum could be spread out over more than a metre. The identification of phytochrome by the same group, in 1959, was made possible by a collaboration between physicists and biologists, and by the development of a new *in vivo* spectrophotometer. The later work on phytochrome physiology could only have been done with a device such as the Ratiospect, a dual-wavelength photometer which allowed the estimation of Pr and Pfr levels, in intact non-green tissues, to be made. Flash photolysis and circular dichroism techniques and, most recently, Spruit's quasi-continuous recording *in vivo* spectrophotometer, made possible the detection of the many intermediates that occur between Pr and Pfr and *vice versa*. During all this time, a remarkable association between biologists and physicists has existed which has been vital to the progress of discovery. It seems inevitable that future research will continue to depend, not only on the collaboration of experts from different disciplines (including, hopefully, biochemists who have not shown much interest as yet), but also on the elaboration of new and even more powerful techniques. Some areas which seem to this author as being ripe for intensive investigation are listed below:

(a) *In vitro* phytochrome effects
The nature of the initial light-mediated changes in membrane properties will probably only be discovered when they can be studied in a cell-free system. A search for *in vitro* phytochrome effects should therefore prove very valuable. What is required is a membrane preparation in which either its physical, chemical, or enzymatic properties can be reversibly modified by red and far-red light. At present, the isolation of plant membranes in sufficient quantities is a crude and relatively unrewarding endeavour relying on homogenization followed by differential centrifugation. It is likely that new techniques for isolating and characterizing membranes will be necessary before any phytochrome-mediated changes *in vitro* will be detectable. One possibility is to use enzymes to remove the cellulose walls followed by hypotonic shock to burst the protoplasts. It may then be possible to separate the plasmalemmas from the contents by sucrose gradient centrifugation. It would also be sensible to 'hedge our bets' by investigating microtubules and microfilaments in more detail.

(b) The importance of intermediates
The mechanism of action of phytochrome under continuous irradiation appears to depend on the simultaneous absorption of Pr and Pfr, and the consequent cycling between the two forms. Under these conditions, the intermediates which have slow dark reactions accumulate, and the level to which they accumulate depends on irradiance. It is clearly very important to obtain more information on this relationship

and to attempt to determine whether the action of phytochrome may be due to the properties of one or more of the intermediates, and not to those of Pfr or Pr. Thus, a detailed analysis of the relationship of intermediate levels to a final morphogenic response (e.g., hypocotyl growth inhibition) needs to be carried out. Admittedly, this is a very difficult project, since only one laboratory currently has the equipment to carry it out (i.e., Spruit's laboratory at Wageningen), and even with this equipment it is not easy to determine precisely how much of the total phytochrome is present as intermediates. Nevertheless, the question is vital to our ideas on phytochrome action, and it is to be hoped that great priority will be given to it, and also to developing similar spectrophotometers in other interested laboratories.

(c) Localization of phytochrome

The precise location of phytochrome in the higher plant cell is still unresolved and intensive efforts to pin it down are to be expected in the near future. It seems unlikely that microbeam *in vivo* spectrophotometers will be able to detect phytochrome in single cells because of the very low levels expected. Even so, it is worth trying since local concentrations of phytochrome, in membranes for example, might result in a much greater absorption than predicted for phytochrome randomly distributed throughout the whole cell. Techniques such as those suggested above for preparing plasmalemma fractions could be used with a dual-wavelength photometer, or a spectrophotometer, to determine whether or not phytochrome is present in those membranes. Finally, the immunological approach already used may be amenable to refinement to provide more precise evidence on cellular location.

(d) The nature of the destruction process

One of the enigmas of phytochrome is the loss of spectral photoreversibility seen when phytochrome is present *in vivo* in the Pfr state. This 'destruction' could be due to any of a number of changes ranging from a change in the environment of the chromophore to complete degradation of the whole molecule. It is most unsatisfactory that we know virtually nothing about the nature of this process. The use of immunology to measure total amount of phytochrome protein before and after a period of destruction would at least tell us whether protein degradation was involved.

(e) Early biochemical effects of light

As mentioned above, the processes occurring in the first hour after light treatment are virtually unknown. An intensive biochemical study of this period would probably pay handsome dividends, although it is difficult to know exactly where to start. Clearly, the reported observations of changes in ATP and NADP concentrations, and in gibberellin levels, within this time period, provide starting points, but a more comprehensive attack would be needed to flush out other changes that might be occurring. Perhaps a survey of other coenzyme levels would be useful – but who really knows?

(f) The control of gene expression

As already mentioned in this chapter, this is probably the area of greatest importance to our general understanding of the mechanisms of development. As far as the photomorphogenic control of gene expression is concerned, the majority of the work that

has been done up to now has been done in a very small numbe[r of labora]tories, and in only one or two test species. Consequently, there i[s much]
independent corroborative studies to be undertaken by other wor[kers with other]
plant species. The principal question which must be solved befor[e an under]standing can be approached is whether the light-mediated increase[in enzyme]
is due to activation of pre-existing inactive enzyme molecules, or t[o their]
de novo synthesis. It is to be hoped that all three of the currently ava[ilable methods]
for assessing the involvement of *de novo* synthesis (i.e., density-lab[elling, radio]
labelling and immunology) will be used in preference to the dubious use [of inhibitors]
of protein and nucleic acid synthesis. If it is found that enzyme synthesis [is controlled,]
the next step will be to determine whether the control is exerted at transc[ription or]
translation, a daunting task indeed. Scientists, however, are incurably op[timistic,]
and one hopes that the enormous amount of effort currently being put into th[e] study
of nucleic acid synthesis will yield, within the foreseeable future, a reliable method of
separating active genes and their primary products, the messenger-RNAs, from the
rest. When this happens a real revolution, not only in photomorphogenesis, but in
the whole of biology, can be expected.

9.3 Why do plants have phytochrome?

A topic which has only been lightly touched upon in this book is the adaptive advantage of the phytochrome system to the higher plant. In chapter 7 the possibility that
phytochrome enables a plant to detect shading from other plants and to respond
accordingly was elaborated. When we consider the higher plants only, this may well
seem to be of high adaptive value, and we might suppose the selection of the phytochrome system through evolution was due to the adaptive advantage thus bestowed.
Phytochrome, however, is present throughout the whole of the eukaryotic green
plant kingdom, and it is difficult to see how this response mechanism to mutual
shading could benefit single-cell algae living in an aquatic environment. On the other
hand, it is intriguing that the only group of plants which do not have phytochrome
are the fungi, which also do not have chlorophyll and do not depend on light for their
existence. Phytochrome, therefore, is associated only with plants which depend on
the light environment, although that statement does not really mean a great deal.

   The initial appearance of the phytochrome system may have occurred when the
progenitors of our modern flora, presumably algae, were living in very different
environmental conditions. Thus, the selection pressures which led to the evolution
of the complex red/far-red photoreversible system may well have disappeared, and
the present functions of phytochrome could represent modifications of its original
function. The environmental conditions leading to the appearance of phytochrome
are now lost in the distant past, and thus it can only be an academic exercise to
attempt to work out what they were. Of all the questions raised in this book, this
last question may be the only one which is truly unanswerable.

Appendix

Light sources for photomorphogenesis experiments

This appendix is provided to assist those interested in carrying out experimental work, either for research or teaching purposes, in the general area of photomorphogenesis. In such experiments it is essential to have complete control over the light sources used.

(a) The 'dark-room'

The first priority is to ensure that a truly dark 'dark-room' is available which can be used both for the growing of plant material and for the experiments themselves. It is essential that the dark-room is relatively large and has at least one good-sized work bench. An ideal arrangement is to have a suite of interconnecting dark-rooms so that the growth areas and the experimental areas can be kept separate. The rooms should be fitted with controlled temperature facilities and preferably with a source of water for watering the plants. These conditions are obviously ideal for the growth of fungi, and since the rooms are almost always dark, fungi can proliferate unseen. Regular inspection should therefore be carried out, together with a scrupulous regard for cleanliness; in particular, unwanted plants should be removed immediately.

The dark-room must have a light-lock so that persons may enter without allowing light to penetrate the growing areas. This is best provided by using a lobby large enough to contain a person, plus room for trolleys, trays, apparatus, etc., which may need to be taken into the room. Alternatively, if space or funds are limited, a heavy black curtain hanging on a semi-circular track outside the dark-room, with sufficient space for a person to enter and open the door in darkness, is adequate if well made.

(b) Light sources

The light sources are undoubtedly the most important pieces of apparatus in photomorphogenesis experiments, and great care should be taken to ensure that only the required wavelengths are being produced. Many different source and filter combinations are possible and the ideal procedure is to measure the spectral distribution of the emitted radiation with a sensitive spectroradiometer. Such devices are, however, very expensive and most laboratories will be unable to go to these lengths. The combinations of sources and filters recommended here will be found adequate for most purposes and the actual spectral distributions should not vary by more than a few per cent from those given in the accompanying figures.

1. *Safelights*. It is necessary to have some light in the dark-room to carry out various operations, and for these purposes a safelight is chosen which is very inefficient in

causing photomorphogenic phenomena. Green light is used, nowadays, in all laboratories. Even green light, however, will set up a photostationary state of about 50 per cent Pfr/P total (see Fig. 4.11), although achieving the photostationary state takes a long time with low irradiances. No light can therefore be truly 'safe' and dark-room lights should be used as sparingly as possible. On no account leave the safelight on for long periods when no one is working in the room.

i. The simplest safelight
Source: Daylight fluorescent tube.
Filter: Three layers (or more) of No. 39 Cinemoid.
(See end of appendix for suppliers.)

This simple safelight is easy to make, and is quite adequate, but transmits some far-red (see Fig. A.1). Most of the far-red light from fluorescent sources, however, comes from the filaments at the ends of the tubes, and can be minimized by wrapping black electrician's tape around the ends.

ii. A better safelight
Source: Daylight or green fluorescent tube.
Filter: One layer of blue Plexiglas (non-far-red transmitting, No. 2045 in the USA or No. 0248 in Europe) and either one layer of amber Plexiglas 245 USA or one layer No. 46 Cinemoid. (See end of appendix, however, for availability of blue Plexiglas.)

If amber Plexiglas is used it must be placed nearest to the source since it fluoresces red under blue light. *Never* use No. 39 Cinemoid with a torch or flashlight. Torches have tungsten bulbs which emit light rich in far-red which is transmitted by the green Cinemoid. To make a portable safelight with a torch use the blue Plexiglas described above with No. 46 Cinemoid.

2. *Irradiation sources.* A general design for all irradiation sources is given in Fig. A.2. It consists simply of two boxes, one above the other with the filter assembly placed in between. Each box should have a door for access to the lights (upper box), and to the plants (lower box). For short irradiation treatments (less than 3–5 minutes) it is probably not necessary to have any cooling devices other than a louvred baffle on the upper chamber. A water bath, as a heat filter, is useful especially for far-red light sources. For longer treatments, steps must be taken to prevent over-heating of either chamber to prevent damage to lamps or plants. This may be readily achieved by using electric fans to blow cold air through each chamber. For very long irradiation treatments (several hours), especially with far-red sources, more sophisticated air-conditioning devices are usually necessary to maintain constant conditions within the irradiation chamber.

In all irradiation chambers care should be taken to prevent light escaping into the rest of the dark-room and exposing dark-grown plants. On the other hand, the irradiation chambers can be kept in a separate interconnected dark-room, but even here it is wise to restrict light emission as much as possible. Fluorescence within the chambers may be a problem (especially with blue light). For this reason it is best to line the chambers with aluminium foil rather than paint, since most paints fluoresce strongly. Similarly, fluorescent materials such as plastic tubing should be masked.

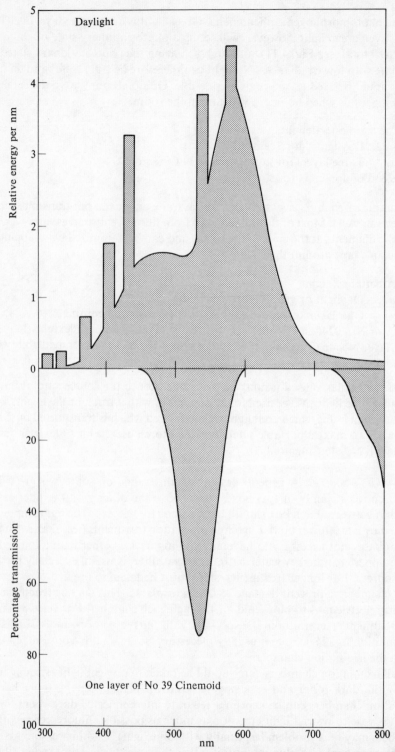

A.1 *Green safelight:* the upper diagram shows the spectral distribution of the radiant energy from a Daylight fluorescent tube, while the lower curve shows the transmission spectrum of the recommended filter.

216   Phytochrome and photomorphogenesis

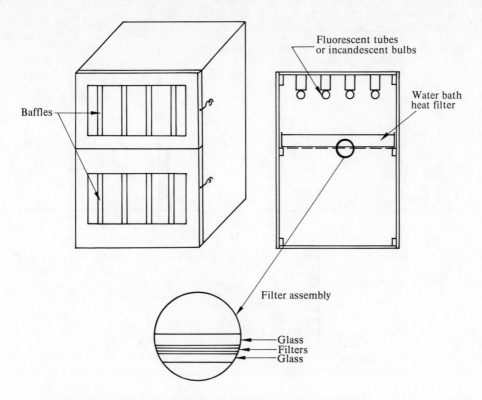

Baffles

Fluorescent tubes
or incandescent bulbs

Water bath
heat filter

Filter assembly

Glass
Filters
Glass

A.2 A simple irradiation chamber suitable for the red, far-red and blue sources described in the text. Dimensions can be altered at will to suit the size of irradiation field required.

3. *Red light.*

  i. The simplest and best assembly
  Source: De Luxe Natural Fluorescent tubes.
  Filter: One layer No. 1 and One layer No. 14 Cinemoid.
  (Spectral distribution in Fig. A.3.)

Note that some red fluorescent tubes have more far-red which is transmitted by the above filter combination.

4. *Far-red light.*

  i. The simplest assembly
  Source: Tungsten Incandescent lamps.
  Filter: Water (10 cm), one layer 5A and one layer 20 Cinemoid.
  (Spectral distribution in Fig. A.4.)

This source works perfectly well using 5×150 W bulbs in a chamber about 0.7 m $\times$ 0.3 m $\times$ 1 m high. It has very little emission below 730 nm, however, and for the purist a better combination is given below.

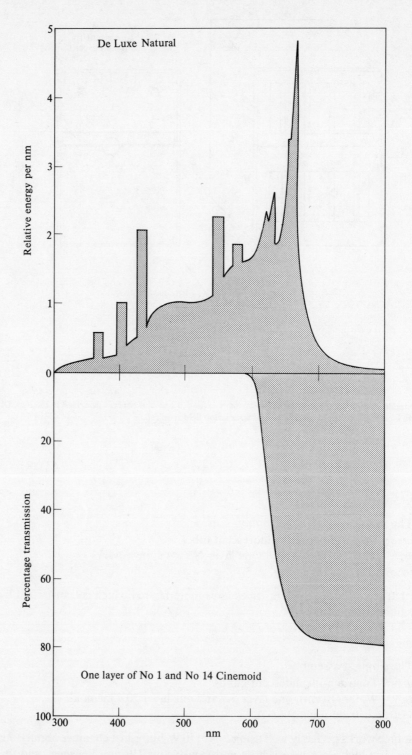

A.3 *Red light:* upper diagram, spectral energy distibution of De Luxe Natural fluorescent tubes; lower diagram, transmission spectrum of the recommended filter combination.

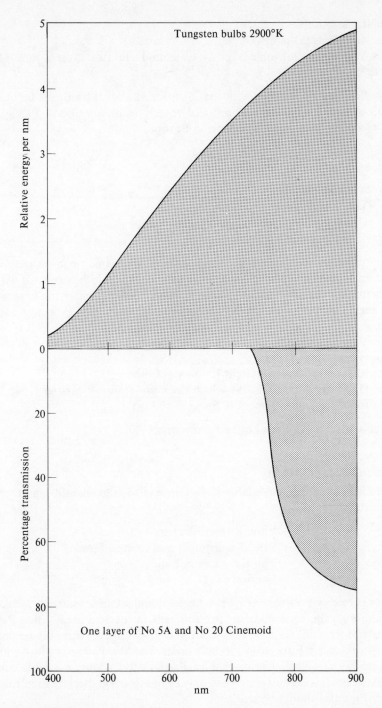

A.4 *Far-red light:* upper diagram, spectral energy distribution of tungsten bulbs; lower diagram, transmission spectrum of the recommended filter combination.

ii. A better assembly
Source: Incandescent lamps.
Filter: Water (10 cm), one layer 14 Cinemoid and one layer 3 mm 627 blue Plexiglas.

This blue Plexiglas (in contrast with that mentioned above) transmits far-red light. In the USA far-red light can be produced readily using only black Plexiglas No. V-58015 as a filter. This is unobtainable in Europe.

5. *Blue light.*

i. The simplest assembly
Source: Tropical Daylight Fluorescent tubes.
Filter: One layer 20 Cinemoid.
(Spectral distribution in Fig. A.5.)

This filter transmits the far-red in the tubes (much of which can be removed by taping the ends), and can give a lower Pfr/Ptotal than would be expected with blue light. This may be very important in long-term experiments where the blue and far-red high-energy reactions might be operating simultaneously. A better source is given below.

ii. A better assembly
Source: Blue or Tropical Daylight Fluorescent tube.
Filter: One layer blue Plexiglas which does not transmit far-red (2045, USA; 0248, Europe: see end of appendix for availability).

This source sets up a Pfr/Ptotal of approximately 0·40.

(c) Filters and their availability

Cinemoid filters are readily available in Britain and their availability in other areas can be ascertained from:

Rank Strand Electrics Ltd.
Sales Department and Central Stores
250, Kennington Lane
London SE 11.

Plexiglas is readily available in North America and some colours are available in Europe, although the code numbers are different in the two areas. Blue Plexiglas which does not transmit far-red (2045 USA and 0248 Europe) can only be obtained on special order, and in Europe only to bulk order. The blue Plexiglas which transmits far-red is readily available. The black Plexiglas V-58015 is only available currently in the USA, where it is expensive and can only be obtained on special order. Enquiries concerning Plexiglas should be sent to:

Rohm and Haas Co.
Philadelphia, 19105
Pa.
U.S.A.

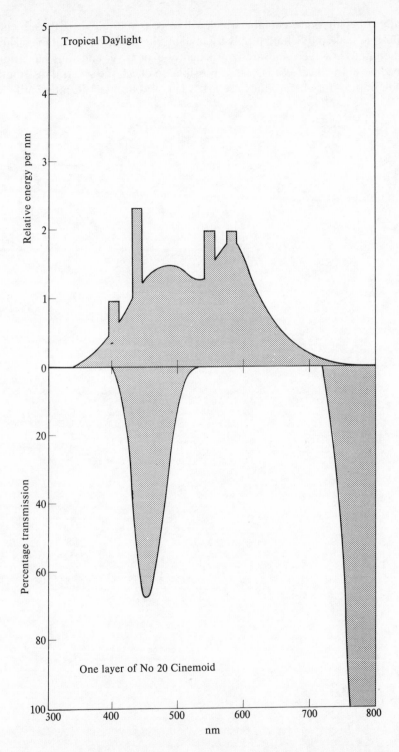

A.5 *Blue light:* upper diagram, spectral energy distribution of a Tropical Daylight fluorescent tube; lower diagram, transmission spectrum of blue Cinemoid (note the high transmission in the far-red region).

Finally, other filters are available which will serve as alternatives to those mentioned here. Both Kodak and Schott supply a range of filters which are suitable. These are glass mounted, however, and are consequently expensive, and thus have not been described here. In all cases it is best, if possible, to check the spectral distribution of all irradiation sources before they are first used, and at intervals afterwards.

References

1. ACTON, G. J. (1972) *Nature New Biology*, **236**, 255.
2. AHMED, S. I. and SWAIN, T. (1970) *Phytochemistry*, **9**, 2287.
3. AMRHEIN, H. and ZENK, M. H. (1971) *Z. Pflanzenphysiol.*, **64**, 145.
4. ANDERSON, G. R., JENNER, E. L. and MUMFORD, F. E. (1969) *Biochemistry*, **8**, 1182.
5. ASOMANING, E. J. A. and GALSTON, A. W. (1961) *Pl. Physiol.*, **36**, 453.
5a. ARNON, D. I., TSUJIMOTO, H. Y. and MCSWAIN, B. D. (1967) *Nature*, **214**, 562.
6. ATTRIDGE, T. H., JOHNSON, C. B. and SMITH, H. (1974). *Biochim. Biophys. Acta,* **343**, 440.
7. ATTRIDGE, T. H. and SMITH, H. (1967) *Biochim. Biophys. Acta*, **148**, 805.
8. ATTRIDGE, T. H. and SMITH, H. (1973) *Phytochemistry*, **12**, 1569.
9. ATTRIDGE, T. H., STEWART, G. R. and SMITH, H. (1972) *F.E.B.S. Letters*, **17**, 84.
10. AUGUSTINUSSEN, E. and MADSEN, A. (1965) *Physiol. Plant.*, **18**, 828.
11. BAUER, L. and MOHR, H. (1959) *Planta*, **54**, 68.
12. BEEVERS, L., LOVEYS, B., PEARSON, J. A. and WAREING, P. F. (1970) *Planta*, **90**, 286.
13. BELLINI, E. and VAN POUCKE, M. (1970) *Planta*, **93**, 60.
14. BERRY, D. R. and SMITH, H. (1971) *J. Cell. Sci.*, **8**, 185.
15. BERTSCH, W. F. (1963) *Am. J. Botany*, **50**, 213.
16. BERTSCH, W. F. and HILLMAN, W. S. (1961) *Am. J. Botany*, **48**, 504.
17. BEWLEY, J. D., BLACK, M. and NEGBI, M. (1967) *Nature*, **215**, 648.
18. BEWLEY, J. D., NEGBI, M. and BLACK, M. (1968) *Planta*, **78**, 351.
19. BICKFORD, E. D. and DUNN, S. (1972) *Lighting for Plant Growth*, Kent State University Press.
20. BIENGER, I. and SCHOPFER, P. (1970) *Planta*, **93**, 152.
21. BIRTH, G. S. (1960) *Agr. Eng.*, **41**, 432.
22. BLAAUW, A. H. (1915) *Z. J. Bot.*, **7**, 465.
23. BLAAUW, O. H. (1963) *Acta Botan. Neerl.*, **10**, 397.
24. BLAAUW-JANSEN, G. (1958) *Acta Botan. Neerl.*, **8**, 1.
25. BLACK, M. (1969) *Symp. Soc. Exptl. Biol.*, **23**, 193.
26. BLACK, M. and WAREING, P. F. (1954) *Nature*, **174**, 705.
27. BLACK, M. and WAREING, P. F. (1958) *Nature*, **181**, 1420.
28. BOISARD, J. (1969) *Physiol. Veg.*, **7**, 119.
29. BOISARD, J., MARME, D. and SCHAFER, E. (1971) *Planta*, **99**, 302.
30. BOISARD, J., SPRUIT, C. J. P. and ROLLIN, P. (1968) *Meded. Landbouwhogesch. Wageningen*, **68**, 1.
31. BOND, G. (1935) *Roy. Soc. Edinburgh Trans.*, **58**, 409.
32. BONNET, C. (1754) *Usage des Feuilles* p. 254.
33. BORTHWICK, H. A. (1957) *Ohio J. Sci.*, **57**, 357.
34. BORTHWICK, H. A. (1972) History of phytochrome. In *Phytochrome*, Mitrakos, K. and Shropshire, W., Jnr. pp. 3–22. Academic Press: London.
35. BORTHWICK, H. A., HENDRICKS, S. B. and PARKER, M. W. (1951) *Bot. Gaz.*, **113**, 95.
36. BORTHWICK, H. A., HENDRICKS, S. B. and PARKER, M. W. (1952) *Proc. Nat. Acad. Sci. U.S.*, **38**, 929.
37. BORTHWICK, H. A., HENDRICKS, S. B., PARKER, M. W., TOOLE, E. H. and TOOLE, V. K. (1952) *Proc. Nat. Acad. Sci. U.S.*, **38**, 662.

38. BORTHWICK, H. A., HENDRICKS, S. B., SCHNEIDER, M. J., TAYLORSON, R. B. and TOOLE, V. K. (1969) *Proc. Nat. Acad. Sci. U.S.*, **64**, 479.

39. BORTHWICK, H. A., HENDRICKS, S. B., TOOLE, E. H. and TOOLE, V. K. (1954) *Bot. Gaz.*, **115**, 205.

40. BORTHWICK, H. A., PARKER, M. W. and HENDRICKS, S. B. (1950) *Am. Soc. Natur.*, **84**, 117.

41. BORTHWICK, H. A., TOOLE, E. H. and TOOLE, V. K. (1964) *Israel J. Botany*, **13**, 112.

42. BOTTOMLEY, W. (1970) *Pl. Physiol.*, **45**, 608.

43. BOTTOMLEY, W., SMITH, H. and GALSTON, A. W. (1966) *Phytochemistry*, **5**, 117.

44. BOULTER, D., ELLIS, R. J. and YARWOOD, A. (1972) *Biol. Rev.*, **47**, 113.

45. BRADBEER, J. W. (1973) In *Biosynthesis and its Regulation in Plants*, ed. Milborrow, B. V., p. 279. Academic Press: London.

46. BRIGGS, W. R. (1963) *Am. J. Botany*, **50**, 196.

47. BRIGGS, W. R. (1963) *Ann. Rev. Pl. Physiol.*, **14**, 311.

48. BRIGGS, W. R. (1963) *Amer. J. Bot.*, **50**, 196.

49. BRIGGS, W. R. (1964) In *Photophysiology*, ed. Giese, A. C., Vol. I, Academic Press: New York.

50. BRIGGS, W. R. and CHON, H. P. (1966) *Pl. Physiol.*, **41**, 1159.

51. BRIGGS, W. R. and FORK, D. C. (1969) *Pl. Physiol.*, **44**, 1081.

52. BRIGGS, W. R. and FORK, D. C. (1969) *Pl. Physiol.*, **44**, 1089.

53. BRIGGS, W. R., GARDNER, G. and HOPKINS, D. W. (1972) In *Phytochrome*, ed. Mitrakos, K. and Shropshire, W. Jnr. p. 145. Academic Press: London.

54. BRIGGS, W. R. and RICE, H. V. (1972) *Ann. Rev. Pl. Physiol.*, **23**, 293.

55. BRIGGS, W. R. and SIEGELMAN, H. W. (1965) *Pl. Physiol.*, **40**, 934.

56. BRIGGS, W. R., TOCHER, R. D. and WILSON, J. F. (1957) *Science*, **126**, 210.

57. BRIGGS, W. R., ZOLLINGER, W. D. and PLATZ, B. B. (1968) *Pl. Physiol.*, **43**, 1239.

58. BROWN, P. K. (1972) *Nature New Biol.*, **236**, 35.

59. BURKE, M. J., PRATT, D. C. and MOSCOWITZ, G. (1972) *Biochemistry*, **11**, 4025.

60. BUTLER, G. L. and BENNETT, V. (1969) *Pl. Physiol.*, **44**, 1285.

61. BUTLER, R. D. and LANE, G. R. (1959) *Journ. Linn. Soc. London (Bot.)*, **56**, 170.

62. BUTLER, W. L., HENDRICKS, S. B. and SIEGELMAN, H. W. (1964) *Photochem. Photobiol.*, **3**, 521.

63. BUTLER, W. L. and LANE, A. C. (1965) *Pl. Physiol.*, **40**, 13.

64. BUTLER, W. L., LANE, H. C. and SIEGELMAN, H. W. (1963) *Pl. Physiol.*, **38**, 514.

65. BUTLER, W. L., NORRIS, K. H., SIEGELMAN, H. W. and HENDRICKS, S. B. (1959) *Proc. Nat. Acad. Sci. U.S.*, **45**, 1703

66. BUTLER, W. L., SIEGELMAN, H. W. and MILLER, C. O. (1964) *Biochemistry*, **3**, 851.

67. CANHAM, A. E. (1966) *Artificial Light in Horticulture*, Centrex Pub. Co.: Eindhoven.

68. CASPARY, R. (1860) *Schr. Kgl. physik okonom Ges. Konigsberg*, **1**, 66.

69. CASTLES, J. J., ROLLESTON, F. S. and WOOD, J. G. (1971) *J. Biol. Chem.*, **246**, 1799.

70. CHOLODNY, N. (1927) *Biol. Zentrabl.*, **47**, 604.

71. CHON, H. P. and BRIGGS, W. R. (1966) *Pl. Physiol.*, **41**, 1715.

72. CLARKSON, D. T. and HILLMAN, W. S. (1967) *Planta*, **75**, 286.

73. COHEN, R. Z. and GOODWIN, T. W. (1962) *Phytochemistry*, **1**, 67.

74. CONE, R. A. (1972) *Nature New Biol.*, **236**, 39.

75. CORRELL, D. L., EDWARDS, J. L., KLEIN, W. H. and SHROPSHIRE, W. Jnr. (1968) *Biochim. Biophys. Acta*, **168**, 36.

76. CORRELL, D. L., EDWARDS, J. L. and SHROPSHIRE, W. Jnr. (1968) *Photochem. Photobiol.*, **8**, 465.

77. CREASY, L. L. (1968) *Phytochemistry*, **7**, 441.

78. CROSS, D. R., LINSCHITZ, H., KASCHE, V. and TENENBAUM, J. (1968) *Proc. Nat. Acad. Sci. U.S.*, **61**, 1095.

79. CUMMING, B. G. (1963) *Can. J. Botany*, **41**, 901.

80. CUMMING, B. G., HENDRICKS, S. B. and BORTHWICK, H. A. (1965) *Can. J. Botany*, **43**, 825.

81. CURRY, G. M. (1957) Ph.D. Thesis, Harvard University (see reference 49).

82. CURRY, G. M. (1969) In *The Physiology of Plant Growth and Development*, ed. Wilkins, M. W. p. 245. McGraw-Hill: Maidenhead.

83. DARWIN, C. (1880) *The Power of Movement in Plants* (assisted by F. Darwin), John Murray: London.
84. DEDEKEN-GRENSON, M. (1954) *Biochim. Biophys. Acta*, **14**, 203.
85. DE LA FUENTE, R. K. and LEOPOLD, A. C. (1968) *Pl. Physiol.*, **43**, 1031.
86. DIEMER, R. (1961) *Planta*, **57**, 111.
87. DITTES, H. and MOHR, H. (1970) *Z. Naturforsch.*, **25b**, 708.
88. DOOSKIN, R. H. and MANCINELLI, A. L. (1968) *Bull. Torrey Bot. Club*, **95**, 474.
89. DOWNS, R. J. (1955) *Pl. Physiol.*, **30**, 468.
90. DOWNS, R. J. (1956) *Pl. Physiol.*, **31**, 279.
91. DOWNS, R. J., HENDRICKS, S. B. and BORTHWICK, H. A. (1957) *Bot. Gaz.*, **118**, 199.
92. DOWNS, R. J. and SIEGELMAN, H. W. (1963) *Pl. Physiol.*, **38**, 25.
93. DRUMM, H., BRUNING, K. and MOHR, H. (1972) *Planta*, **106**, 259.
94. DRUMM, H., EICHINGER, I., MOLLER, J., PETER, K. and MOHR, H. (1971) *Planta*, **99**, 265.
95. DRUMM, H., FALK, H., MOLLER, J. and MOHR, H. (1970) *Cytobiologie*, **2**, 335.
96. DURST, F. and DURANTON, H. (1970) *C. R. Acad. Sci.*, **270**, 2940.
97. DURST, F. and MOHR, H. (1966) *Naturwissenschaften*, **53**, 531.
98. DURST, F. and MOHR, H. (1966) *Naturwissenschaften*, **53**, 707.
99. EBERHARDT, F. (1954) *Planta*, **43**, 253.
100. ELLIS, R. J. and MACDONALD, I. R. (1970) *Pl. Physiol.*, **46**, 227.
101. ENGELSMA, G. (1967) *Planta*, **75**, 207.
102. ENGLESMA, G. (1967) *Naturwissenschaften*, **54**, 319.
103. ENGELSMA, G. (1968) *Planta*, **82**, 355.
104. ENGELSMA, G. (1969) *Naturwissenschaften*, **56**, 563.
105. ENGELSMA, G. (1970) *Acta Bot. Neerl.*, **19**, 403.
106. ENGELSMA, G. and MEIJER, G. (1965) *Acta Bot. Neerl.*, **14**, 54.
107. ETZOLD, H. (1965) *Planta*, **64**, 254.
108. EVANS, L. T., HENDRICKS, S. B. and BORTHWICK, H. A. (1967) *Planta*, **64**, 201.
109. EVENARI, M. (1956) In *Radiation Biology*, ed. Hollaender, A. Vol. 3, p. 519. McGraw-Hill: New York.
110. EVENARI, M. (1965) In *Encyclopaedia of Plant Physiology*, ed. Ruhland, W. Vol. 15 (2). Springer Verlag: Berlin.
111. EVENARI, M. and NEUMANN, G. (1953) *Palest. J. Bot. Jerusalem Ser.*, **6**, 96.
112. EVENARI, M. and NEUMANN, G. (1953) *Bull. Res. Council Israel*, **3**, 136.
113. EVENARI, M., NEUMANN, G. and STEIN, G. (1957) *Nature*, **180**, 609.
114. EVERETT, M. S. and BRIGGS, W. R. (1970) *Pl. Physiol.*, **45**, 679.
115. EVERETT, M. S., BRIGGS, W. R. and PURVES, W. K. (1970) *Pl. Physiol.*, **45**, 805.
116. FEIERABEND, J. and PIRSON, A. (1966) *Z. Pflanzenphysiol.*, **55**, 235.
117. FILNER, B. and KLEIN, A. O. (1969) *Pl. Physiol.*, **43**, 1587.
118. FILNER, P. and VARNER, J. E. (1967) *Proc. Nat. Acad. Sci. U.S.*, **58**, 1520.
119. FLETCHER, R. A. and SALIK, S. (1964) *Pl. Physiol.*, **39**, 328.
120. FLINT, L. H. and MCALISTER, E. D. (1935) *Smithsonian Misc. Coll.*, **94**, 1.
121. FLINT, L. H. and MCALISTER, E. D. (1937) *Smithsonian Misc. Coll.*, **96**, 1.
122. FONDEVILLE, J. C., BORTHWICK, H. A. and HENDRICKS, S. B. (1966) *Planta*, **69**, 357.
122a. FONDEVILLE, J. C., SCHNEIDER, M. J., BORTHWICK, H. A. and HENDRICKS, S. B. (1967) *Planta*, **75**, 228.
123. FRANKLAND, B. (1972) In *Phytochrome*, ed. Mitrakos, K. and Shropshire, W. Jnr. p. 195. Academic Press: London.
124. FRANKLAND, B. and SMITH, H. (1967) *Planta*, **77**, 354.
125. FREDERICQ, H. (1964) *Pl. Physiol.*, **39**, 812.
126. FREDERICQ, H. (1965) *Biol. Jaarboek.*, **33**, 66.
127. FREDERICQ, H. and DE GREEF, J. (1966) *Photochem. Photobiol.*, **5**, 431.
128. FRIEND, D. J. C. (1965) *Can. J. Botany*, **43**, 161.
129. FUJII, R. (1965) *Physiol. Ecol. (Tohoku)*, **7**, 79.
130. FUGII, T. (1962) *Botan. Mag. (Tokyo)*, **75**, 56.
131. FURUYA, M. (1968) In *Progress in Phytochemistry*, eds. Reinhold, L. and Liwschitz, Y. Vol. 1, pp. 347–405. Interscience: London.
132. FURUYA, M., GALSTON, A. W. and STOWE, B. B. (1962) *Nature*, **193**, 456.

133. FURUYA, M. and HILLMAN, W. S. (1964) *Planta*, **63**, 31.
134. FURUYA, M., HOPKINS, W. G. and HILLMAN, W. S. (1965) *Arch. Biochem. Biophys.*, **112**, 180.
135. FURUYA, M. and THOMAS, R. G. (1964) *Pl. Physiol.*, **39**, 634.
136. FURUYA, M. and TORREY, J. G. (1964) *Pl. Physiol.*, **39**, 987.
137. GALSTON, A. W. (1968) *Proc. Nat. Acad. Sci. U.S.*, **61**, 454.
138. GALSTON, A. W. and BAKER, R. S. (1951) *Amer. J. Botany*, **38**, 190.
139. GALSTON, A. W. and KAUR, R. (1961) In *Light and Life*, ed. McElroy, W. D. and Glass, H. B. pp. 687–705. Johns Hopkin University Press: Washington.
140. GALSTON, A. W. and SATTER, R. L. (1972) In *Recent Advances in Phytochemistry*, eds. Runeckles, V. C. and Tso, T. C. Vol. 5, p. 51. Academic Press: New York.
141. GARDNER, G., PIKE, C. S., RICE, H. V. and BRIGGS, W. R. (1971) *Pl. Physiol.*, **48**, 686.
142. GARNER, W. W. and ALLARD, H. A. (1920) *J. Agric. Res.*, **18**, 553.
143. GASSMAN, M., PLUSCEC, J. and BOGORAD, L. (1968) *Pl. Physiol.*, **43**, 1411.
144. GILES, K. L. and VON MALTZAHN, K. E. (1968) *Can. J. Bot.*, **46**, 305.
145. GOESCHL, J. D., PRATT, H. K. and BONNER, B. A. (1967) *Pl. Physiol.*, **42**, 1077.
146. GOODWIN, R. H. (1942) *Am. J. Botany*, **29**, 818.
147. GORDON, S. A. (1961) In *Proc. 3rd Int. Congr. Photobiol.*, ed. Christensen, B. C. and Buchmann, B. p. 441. Elsevier: Amsterdam.
148. GOREN, R. and GALSTON, A. W. (1966) *Pl. Physiol.*, **41**, 1055.
149. GORTER, C. J. (1961) *Pl. Physiol.*, **14**, 332.
150. GRAHAM, D., GRIEVE, A. M. and SMILLIE, R. M. (1968) *Nature*, **218**, 89.
151. GRANICK, S. (1959) *Pl. Physiol.*, **34**, 5.
152. GRANICK, S. (1965) In *Biochemistry of the Chloroplasts*, ed. Goodwin, T. W. Vol. 2, p. 373. Academic Press: London.
153. GRANTLIPP, A. E. and BALLARD, L. A. T. (1963) *Austral. J. Biol.*, **16**, 572.
154. GREGORY, R. P. F. (1971) *Biochemistry of Photosynthesis*, Wiley Interscience: London.
155. GRESSEL, J., GALUN, E. and STRAUSBAUCH, L. (1971) *Nature*, **232**, 648
156. GRILL, R. (1968) *Planta*, **85**, 42.
157. GRILL, R. and VINCE, D. (1965) *Planta*, **67**, 122.
158. GRILL, R. and VINCE, D. (1966) *Planta*, **70**, 1.
159. GUNNING, B. E. S. and JAGOE, M. P. (1965) In *Biochemistry of the Chloroplasts*, ed. Goodwin, T. W. Vol. 2, pp. 655–676. Academic Press: London.
160. HACKER, M., HARTMANN, K. M. and MOHR, H. (1964) *Planta*, **63**, 253.
161. HAHLBROCK, K. and GRISEBACH, H. (1970) *F.E.B.S. Letters*, **11**, 62.
162. HAHLBROCK, K., KUHLEN, E. and LINDL, T. (1971) *Planta*, **99**, 311.
163. HAHLBROCK, K., SUTTER, A., WELLMANN, E., ORTMANN, R. and GRISEBACH, H. (1971) *Phytochemistry*, **10**, 109.
164. HAHLBROCK, K. and WELLMANN, E. (1970) *Planta*, **94**, 236.
165. HAHN, L. W. and MILLER, J. H. (1966) *Physiol. Plant.*, **19**, 134.
166. HARPER, D. B., AUSTIN, D. J. and SMITH, H. (1970) *Phytochemistry*, **9**, 497.
167. HARTMANN, K. M. (1966) *Photochem. Photobiol.*, **5**, 349.
168. HARTMANN, K. M. (1967) *Z. Naturforsch.*, **22b**, 1172.
169. HAUPT, W. (1959) *Planta*, **53**, 484.
170. HAUPT, W. (1970) *Physiol Veg.*, **8**, 551.
171. HAUPT, W. (1970) *Z. Pflanzenphysiol.*, **62**, 287.
172. HAUPT, W. (1972) In *Phytochrome*, ed. Mitrakos, K. and Shropshire, W. Jnr. p. 349. Academic Press: London.
173. HAUPT, W. (1972) In *Phytochrome*, ed. Mitrakos, K. and Shropshire, W. Jnr. p. 553. Academic Press: London.
174. HAUPT, W., MORTEL, G. and WINKELNKEMPER, I. (1969) *Planta*, **88**, 183.
175. HAUPT, W. and SCHONBOHM, E. (1970) In *Photobiology of Microorganisms*, ed. Holldal, P. p. 238. Wiley: Chichester.
176. HAUPT, W. and THIELE, R. (1961) *Planta*, **56**, 388.
177. HAVIR, E. A. and HANSON, K. R. (1968) *Biochemistry*, **7**, 1896.
178. HENDRICKS, S. B. (1959) In *Photoperiodism and Related Phenomena in Plants and Animals*, ed. Withrow, R. B. p. 423. A.A.A.S: Washington.

179. HENDRICKS, S. B. (1960) *Cold Spring Harbour Symp. Quart. Biol.*, **25**, 245.
180. HENDRICKS, S. B. (1968) *Scientific American* (Sept.), 174.
181. HENDRICKS, S. B. and BORTHWICK, H. A. (1959) *Proc. Nat. Acad. Sci. U.S.*, **45**, 344.
182. HENDRICKS, S. B. and BORTHWICK, H. A. (1967) *Proc. Nat. Acad. Sci. U.S.*, **58**, 2125.
183. HENDRICKS, S. B., BUTLER, W. L. and SIEGELMAN, H. W. (1962) *J. Phys. Chem.*, **66**, 2550.
184. HENNINGSEN, K. W. (1965) In *Biochemistry of the Chloroplasts*, ed. Goodwin, T. W. Vol. 2, p. 453. Academic Press: London.
185. HENNINGSEN, K. W. and BOYNTON, J. E. (1969) *J. Cell. Sci.*, **5**, 757.
186. HENSHALL, J. D. and GOODWIN, T. W. (1964) *Photochem. Photobiol.*, **3**, 243.
187. HENSHALL, J. D. and GOODWIN, T. W. (1964) *Phytochemistry*, **3**, 677.
188. HILLMAN, W. S. (1959) In *Photoperiodism and Related Phenomena in Plants and Animals*, ed. Withrow, R. B. pp. 181–196, A.A.A.S.: Washington.
189. HILLMAN, W. S. (1962) *The Physiology of Flowering*, Holt, Rinehart and Winston: New York.
190. HILLMAN, W. S. (1964) *Am. J. Bot.*, **51**, 1102.
191. HILLMAN, W. S. (1965) *Physiol. Plant.*, **18**, 346.
192. HILLMAN, W. S. (1966) *Pl. Physiol.*, **41**, 907
193. HILLMAN, W. S. (1967) *Ann. Rev. Pl. Physiol.*, **18**, 301.
194. HILLMAN, W. S. and GALSTON, A. W. (1957) *Pl. Physiol.*, **32**, 129.
195. HILLMAN, W. S. and KOUKKARI, W. H. (1967) *Pl. Physiol.*, **42**, 1413.
196. HILLMAN, W. S. and PURVES, W. K. (1966) *Planta*, **70**, 275.
197. HOPKINS, D. W. (1971) Ph.D. Thesis Univ. Calif., San Diego (cited in ref. 54).
197a. HOPKINS, D. W. and BRIGGS, W. R. (1973) *Pl. Physiol.*, **51**, 525.
198. HOPKINS, D. W. and BUTLER, W. L. (1970) *Pl. Physiol.*, **45**, 567.
199. HOPKINS, W. G. and HILLMAN, W. S. (1965) *Am. J. Bot.*, **52**, 427.
200. HOPKINS, W. G. and HILLMAN, W. S. (1965) *Planta*, **65**, 157.
201. HOPKINS, W. G. and HILLMAN, W. S. (1966) *Pl. Physiol.*, **41**, 593.
202. HU, A. S. L., BOCK, R. M. and HALVORSON, M. O. (1962) *Anal. Biochem.*, **4**, 489.
203. IKUMA, H. and THIMANN, H. V. (1960) *Pl. Physiol.*, **35**, 557.
204. IKUMA, H. and THIMANN, H. V. (1964) *Pl. Physiol.*, **39**, 756.
205. IREDALE, S. E. and SMITH, H. (1973) *Phytochemistry*, **12**, 2145.
206. ISIKAWA, S. (1954) *Botan. Mag., Tokyo*, **67**, 51.
207. ISIKAWA, S. (1957) *Botan. Mag., Tokyo*, **70**, 264.
208. ISIKAWA, S. and OOHUSA, T. (1956) *Botan. Mag., Tokyo*, **69**, 132.
209. ISIKAWA, S. and SHIMOGAWANA, G. (1954) *J. Jap. Forestry Soc.*, **36**, 317.
210. JAFFE, M. J. (1968) *Science*, **162**, 1016.
211. JAFFE, M. J. (1970) *Pl. Physiol.*, **46**, 768.
212. JAFFE, M. J. (1972) In *Recent Advances in Phytochemistry*, ed. Runeckles, V. C. and Tso, T. C. Vol. 5, p. 80. Academic Press: New York.
213. JAFFE, M. J. and GALSTON, A. W. (1967) *Planta*, **77**, 135.
214. JOHNSON, C. B., ATTRIDGE, T. H. and SMITH, H. (1973) *Biochim. Biophys. Acta*, **317**, 219.
215. JONES, M. B. and BAILEY, L. F. (1956) *Pl. Physiol.*, **31**, 347.
216. JONES, R. W. and SHEARD, R. W. (1972) *Nature New Biol.*, **238**, 222.
217. JOST, J. P. and RICKENBERG, H. V. (1971) *Ann. Rev. Biochem.*, **41**, 741.
218. KADMAN-ZAHAVI, A. (1960) *Bull. Res. Comm. Israel*, **9**, D-1.
219. KAHN, A. (1960) *Pl. Physiol.*, **35**, 1.
220. KAHN, A. (quoted by Virgin, ref. 465).
221. KAHN, A. (1968) *Pl. Physiol.*, **43**, 1781.
222. KAHN, A., GOSS, J. A. and SMITH, D. E. (1957) *Science*, **125**, 645.
223. KANDELER, R. (1963) *Naturwissenschaften*, **50**, 551.
224. KANG, B. G. and RAY, P. M. (1969) *Planta*, **87**, 193.
225. KANG, B. G. and RAY, P. M. (1969) *Planta*, **87**, 206.
226. KANG, B. G. and RAY, P. M. (1969) *Planta*, **87**, 217.
227. KAROW, H. and MOHR, H. (1967) *Planta*, **72**, 170.
228. KASEMIR, H. and MOHR, H. (1972) *Pl. Physiol.*, **49**, 453.
229. KENDE, H. and LANG, A. (1964) *Pl. Physiol.*, **39**, 435.
230. KENDRICK, R. E. and FRANKLAND, B. (1968) *Planta*, **82**, 317.

231. KENDRICK, R. E. and SPRUIT, C. J. P. (1972) *Nature New Biol.*, **237**, 281.
232. KENDRICK, R. E. and SPRUIT, C. J. P. (1973) *Photochem. Photobiol.*, **18**, 139.
234. KENDRICK, R. E. and SPRUIT, C. J. P. (1973) *Photochem Photobiol.*, **18**, 153.
235. KENDRICK, R. E., SPRUIT, C. J. P. and FRANKLAND, B. (1969) *Planta*, **88**, 293.
236. KINZEL, W. (1908) *Ber. dt. bot. Ges.*, **26a**, 105.
237. KIRK, J. T. O. (1970) *Ann. Rev. P. Physiol.*, **21**, 11.
238. KIRK, J. T. O. and ALLEN, R. L. (1965) *Biochem. Biophys. Res. Commun.*, **21**, 523.
239. KIRK, J. T. O. and TILNEY-BASSETT, J. A. E. (1967) *The Plastids*, Freeman: London.
240. KLEIBER, H. and MOHR, H. (1967) *Planta*, **76**, 85.
241. KLEIN, R. M. (1965) *Physiol. Plant*, **18**, 1026.
242. KLEIN, R. M. and EDSALL, P. C. (1966) *Pl. Physiol.*, **41**, 949.
243. KLEIN, S., BRYAN, G. and BOGORAD, L. (1964) *J. Cell. Biol.*, **22**, 433.
244. KLEIN, W. H. (1959) In *Photoperiodism and Related Phenomena in Plants and Animals*, ed. Withrow, R. B. pp. 207–215. A.A.A.S.: Washington.
245. KLEIN, W. H., PRICE, L. and MITRAKOS, K. (1963) *Photochem. Photobiol.*, **2**, 233.
246. KLEIN, W. H., WITHROW, R. B. and ELSTAD, V. B. (1956) *Pl. Physiol.*, **31**, 289.
247. KLEIN, W. H., WITHROW, R. B., ELSTAD, V. B. and PRICE, L. (1957) *Am. J. Botany*, **44**, 15.
248. KOHLER, D. (1969) *Planta*, **84**, 158.
249. KOLLER, D. and NEGBI, M. (1959) *Ecology*, **40**, 20.
250. KOLLER, B. and SMITH, H. (1972) *Phytochemistry*, **11**, 1295.
251. KOUKKARI, W. L. and HILLMAN, W. S. (1966) *Physiol. Plant.*, **19**, 1073.
252. KOUKKARI, W. L. and HILLMAN, W. S. (1968) *Pl. Physiol.*, **43**, 698.
253. KROES, H. H. (1970) *Meded. Landbouwhogesch. Wageningen*, **70**, 18.
254. LANDGRAF, J. E. (1961) *Planta*, **57**, 543.
255. LANE, H. C. and KASPERBAUER, M. J. (1965) *Pl. Physiol.*, **40**, 109.
256. LANGE, H. and MOHR, H. (1965) *Planta*, **67**, 107.
257. LANGE, S. (1927) *Jahrb. Wiss. Bot.*, **67**, 1.
258. LEFF, J. (1964) *Pl. Physiol.*, **39**, 299.
259. LE NOIR, JNR., W. C. (1967) *Am. J. Botany*, **54**, 876.
260. LINSCHITZ, H., KASCHE, V., BUTLER, W. L. and SIEGELMAN, H. W. (1966) *J. biol. Chem.*, **241**, 3395.
261. LIVERMAN, J. L. and BONNER, J. (1953) *Proc. Nat. Acad. Sci. U.S.*, **39**, 905.
262. LIVERMAN, J. L., JOHNSON, M. P. and STARR, L. (1955) *Science*, **121**, 440.
263. LOCKHART, J. A. (1956) *Proc. Nat. Acad. Sci. U.S.*, **42**, 841.
264. LOCKHART, J. A. (1959) *Pl. Physiol.*, **34**, 457.
265. LOCKHART, J. A. (1961) *Am. J. Botany*, **48**, 516.
266. LOCKHART, J. A. and GOTTSCHALL, V. (1959) *Pl. Physiol.*, **34**, 460.
267. LOENING, U. E. (1967) *Biochem. J.*, **102**, 251.
268. LOERCHER, L. (1966) *Pl. Physiol.*, **41**, 932.
269. LOVEYS, B. R. and WAREING, P. F. (1971) *Planta*, **98**, 109.
270. MCCULLOUGH, J. M. and SHROPSHIRE, W. Jnr. (1970) *Pl. Cell Physiol.*, **11**, 139.
271. MCDONOUGH, W. T. (1967) *Nature*, **214**, 1147.
272. MACDOUGAL, D. T. (1903) *N.Y. Bot. Gard. Mem.*, **2**, 319.
273. MAIER, W. (1933) *Jb. wiss. Bot.*, **77**, 321.
274. MANCINELLI, A. L. and BORTHWICK, H. A. (1964) *Annals di Botanica (Milan)*, **28**, 9.
275. MANCINELLI, A. L., BORTHWICK, H. A. and HENDRICKS, S. B. (1966) *Botan. Gaz.*, **121**, 1.
276. MANCINELLI, A. L., YANIV, Z. and SMITH, P. (1967) *Pl. Physiol.*, **42**, 333.
277. MARGULIES, M. M. (1965) *Pl. Physiol.*, **40**, 57.
278. MARGULIES, M. M. (1967) *Pl. Physiol.*, **42**, 218.
279. MARME, D. and SCHAFER, E. (1972) *Z. Pflanzenphysiol.*, **67**, 192.
280. MAPSON, L. W. and SWAIN, T. (1961) *Nature*, **204**, 886.
281. MARSH, H. V., HAVIR, E. A. and HANSON, K. R. (1968) *Biochemistry*, **7**, 1915.
282. MEIJER, G. (1959) *Acta Bot. Neerl.*, **8**, 189.
283. MEIJER, G. (1968) *Acta Bot. Neerl.*, **17**, 9.
284. MEIJER, G. (1971) *Acta Horticulturae*, **22**, 104.
285. MILLER, C. O. (1956) *Pl. Physiol.*, **31**, 318.
286. MILLER, C. O. (1958) *Pl. Physiol.*, **33**, 115.

287. MILLER, D. H. and MACHLIS, L. (1966) *Pl. Physiol.*, **41**, 515.
288. MILLER, J. H. and MILLER, P. H. (1961) *Am. J. Bot.*, **48**, 154.
289. MILLER, J. W. and TENG, D. (1967) *Proc. Int. Congr. Biochem. 7th Tokyo*, 1059.
290. MITRAKOS, K. (1972) In *Phytochrome*, ed. Mitrakos, K. and Shropshire, W. Jnr. p. 587. Academic Press: London.
291. MITRAKOS, K., KLEIN, W. H. and PRICE, L. (1965) *Planta*, **66**, 207.
292. MOHR, H. (1957) *Planta*, **49**, 389.
293. MOHR, H. (1964) *Biol. Rev.*, **39**, 87.
294. MOHR, H. (1966) *Z. Pflanzenphysiol.*, **54**, 63.
295. MOHR, H. (1966) *Photochem. Photobiol.*, **5**, 469.
296. MOHR, H. (1969) In *Physiology of Plant Growth and Development*, ed. Wilkins, M. B. p. 509. McGraw-Hill: London.
297. MOHR, H. (1972) *Lectures in Photomorphogenesis*, Springer: Berlin.
298. MOHR, H. and APPUHN, H. (1962) *Planta*, **59**, 49.
299. MOHR, H. and APPUHN, H. (1963) *Planta*, **60**, 274.
300. MOHR, H. and BIENGER, I. (1967) *Planta*, **75**, 180.
301. MOHR, H., BIENGER, I. and LANGE, H. (1971) *Nature*, **230**, 56.
302. MOHR, H. and HANG, A. (1962) *Planta*, **59**, 151.
303. MOHR, H., MEIER, U. and HARTMANN, K. (1964) *Planta*, **60**, 483.
304. MOHR, H. and PICHLER, I. (1960) *Planta*, **55**, 57.
305. MOHR, H., SCHLICKEWEI, I. and LANGE, H. (1965) *Z. Naturforsch.*, **20**, 819.
306. MOHR, H. and SENF, R. (1966) *Planta*, **71**, 195.
307. MOHR, H. and WEHRUNG, M. (1960) *Planta*, **55**, 438.
308. MUMFORD, F. H. and JENNER, E. L. (1966) *Biochemistry*, **5**, 3657.
309. MUMFORD, F. E., SMITH, D. H. and CASTLE, J. E. (1961) *Pl. Physiol.*, **36**, 752.
310. MUMFORD, F. E., SMITH, D. H. and HEYTLER, A. G. (1964) *Biochem. J.*, **91**, 517.
311. MUSGRAVE, A., KAYS, S. E. and KENDE, H. (1969) *Planta*, **89**, 165.
312. NADLER, K. and GRANICK, S. (1970) *Pl. Physiol.*, **46**, 240.
313. NAKATA, S. and LOCKHART, J. A. (1966) *Am. J. Botany*, **53**, 12.
314. NAKAYAMA, S., BORTHWICK, H. A. and HENDRICKS, S. B. (1960) *Botan. Gaz.*, **121**, 237.
315. NEWMANN, I. A. and BRIGGS, W. R. (1972) *Pl. Physiol.*, **50**, 687.
315a. NEUSCHELER-WIRTH, H. (1970) *Z. Pflanzenphysiol*, **63**, 238.
316. NEYLAND, M., NG, Y. L. and THIMANN, K. V. (1963) *Pl. Physiol.*, **38**, 447.
317. NITSCH, C. and NITSCH, J. P. (1966) *C. R. Acad. Sci. (Paris)*, **262**, 1102.
318. NUTILE, G. E. (1945) *Pl. Physiol.*, **20**, 433.
319. OELZE-KAROW, H. and MOHR, H. (1970) *Z. Naturforsch.*, **25b**, 1282.
320. OELZE-KAROW, H. SCHOPFER, P and MOHR, H. (1970) *Proc. Nat. Acad. Sci. U.S.*, **65**, 51.
321. OLSON, J. S. and NIENSTAEDT, H. (1957) *Science*, **125**, 492.
322. PARKER, M. W., HENDRICKS, S. B. and BORTHWICK, H. A. (1950) *Bot. Gaz.*, **111**, 242.
323. PARKER, M. W., HENDRICKS, S. B., BORTHWICK, H. A. and SCULLY, N. J. (1946) *Botan. Gaz.*, **108**, 1.
324. PARKER, M. W., HENDRICKS, S. B., BORTHWICK, H. A., and WENT, F. W. (1949) *Am. J. Botany*, **36**, 194.
325. PHILLIPS, I. D. J., VLITOS, J. A. J. and CUTLER, H. (1959) *Contribs. Boyce Thompson Inst.*, **20**, 111.
326. PICKARD, B. G. and THIMANN, K. V. (1964) *Pl. Physiol.*, **39**, 341.
327. PIKE, C. S. and BRIGGS, W. R. (1972) *Pl. Physiol.*, **49**, 514.
328. PIKE, C. S. and BRIGGS, W. R. (1972) *Pl. Physiol.*, **49**, 521.
329. PINE, K. and KLEIN, A. O. (1972) *Dev. Biol.*, **28**, 280.
330. PIRINGER, A. A. and HEINZE, P. H. (1954) *Pl. Physiol.*, **29**, 467.
331. PJOH, C. J. and FURUYA, M. (1967) *Plant Cell Physiol.*, **8**, 4.
332. PORTER, D. W. and ANDERSON, D. G. (1962) *Arch. Biochem. Biophys.*, **97**, 520.
333. PORTER, G. (1969) In *An Introduction to Photobiology,* ed. Swanson, C. P. p. 1. Prentice-Hall: New Jersey.
334. POWELL, R. D. (1963) *Am. Biol. Teacher*, **25**, 107.
335. POWELL, R. D. and GRIFFITH, M. M. (1960) *Pl. Physiol.*, **35**, 273.
336. PRATT, L. H. and BRIGGS, W. R. (1966) *Pl. Physiol.*, **41**, 467.

336a. PRATT, L. H. and BUTLER, W. L. (1968) *Photochem. Photobiol.*, **8**, 477.
337. PRATT, L. H. and COLEMAN, R. A. (1971) *Proc. Nat. Acad. Sci. U.S.*, **68**, 2431.
338. PRICE, L., MITRAKOS, K. and KLEIN, W. H. (1964) *Quart. Rev. Biol.*, **39**, 11.
339. PRICE, L., MITRAKOS, K. and KLEIN, W. H. (1965) *Physiol. Plant.*, **18**, 540.
340. PRIESTLEY, J. H. (1925) *New Phytol.*, **24**, 271.
341. PRIESTLEY, J. H. (1926) *New Phytol.*, **25**, 145.
342. PRIESTLEY, J. H. and EWING, J. (1923) *New Phytol.*, **22**, 30.
343. PRUE, D. (in press) *Photoperiodism in Plants*, McGraw-Hill: London.
344. PURVES, W. K. and BRIGGS, W. R. (1968) *Pl. Physiol.*, **43**, 1259.
345. QUAIL, P., SCHAFER, E. and MARME, D. (1972) *Book of Abstracts, VI Int. Photobiology Congress*, Bochum, Germany.
346. QUAIL, P., SCHAFER, E. and MARME, D. (1973) *Nature*, **245**, 189.
347. RABINOWITCH, E. and GOVINDJEE (1969) *Photosynthesis*, Wiley: New York.
348. RACUSEN, R. and MILLER, K. (1972) *Pl. Physiol.*, **49**, 654.
349. RAU, W. (1967) *Planta*, **72**, 14.
350. RAY, J. (1686) *Historia plantarium*, **1**, 15.
351. REID, D. M. and CLEMENTS, J. B. (1968) *Nature*, **219**, 607.
352. REID, D. M., CLEMENTS, J. B. and CARR, D. J. (1968) *Nature*, **217**, 580.
353. REID, D. M., TUING, M. S., DURLEY, R. C. and RAILTON, I. D. (1972) *Planta*, **108**, 67.
353a. RETHY, R. (1968) *Z. Pflanzenphysiol.*, **59**, 100.
354. RICE, H. V. and BRIGGS, W. R. (1973) *Pl. Physiol.*, **51**, 939.
355. RICE, H. V., BRIGGS, W. R. and JACKSON-WHITE, C. R. (1973) *Pl. Physiol.*, **51**, 917.
356. RISSLAND, I. and MOHR, H. (1967) *Planta*, **77**, 239.
357. ROBERTS, D. W. A. and PERKINS, H. J. (1962) *Biochim. Biophys. Acta*, **58**, 499.
358. ROLLIN, P. and MAIGRAN, G. (1967) *Nature*, **214**, 741.
359. ROUX, S. J. (1971) Ph.D. Thesis, Univ. of Yale (cited in ref. 54).
360. ROUX, S. J. and HILLMAN, W. J. (1969) *Arch. Biochem. Biophys.*, **131**, 423.
361. RUBINSTEIN, B., DRURY, K. S. and PARK, R. B. (1969) *Pl. Physiol.*, **44**, 105.
362. RUDIGER, W. (1972) In *Phytochrome*, ed. Mitrakos, K. and Shropshire, W. Jnr. p. 129. Academic Press: London.
363. RUDIGER, W. and CORRELL, D. L. (1969) *Justus Liebigs Ann. Chem.*, **723**, 208.
364. RUSSELL, D. W. (1971) *J. Biol. Chem.*, **246**, 3870.
365. RUSSELL, D. W. and GALSTON, A. W. (1967) *Phytochemistry*, **6**, 791.
366. RUSSELL, D. W. and GALSTON, A. W. (1969) *Pl. Physiol.*, **44**, 1211.
367. SACHER, J. A., TOWERS, G. H. N. and DAVIES, D. D. (1972) *Phytochemistry*, **11**, 2383.
368. SANDMEIER, M. and IVART, J. (1972) *Photochem. Photobiol.*, **16**, 51.
369. SATTER, R. L. and GALSTON, A. W. (1971) *Science*, **174**, 518.
370. SATTER, R. L. and GALSTON, A. W. (1971) *Pl. Physiol.*, **48**, 740.
371. SATTER, R. L., MARINOFF, P. and GALSTON, A. W. (1970) *Am. J. Botany*, **57**, 916.
372. SATTER, R. L., SABNIS, D. D. and GALSTON, A. W. (1970) *Am. J. Botany*, **57**, 374.
373. SCHAFER, E., MARCHAL, B. and MARME, D. (1972) *Photochem. Photobiol.*, **15**, 457.
374. SCHAFER, E., SCHMIDT, J., QUAIL, P. and MARME, D. presented at VIth Int. Photobiology Congress, Bochum, Germany, 1972.
375. SCHARFF, O. (1962) *Physiol. Plant.*, **15**, 804.
376. SCHEIBE, J. and LANG, A. (1965) *Pl. Physiol.*, **40**, 485.
377. SCHERF, H. and ZENK, M. H. (1967) *Z. Pflanzenphysiol.*, **56**, 203.
378. SCHNEIDER, M. J., BORTHWICK, H. A. and HENDRICKS, S. B. (1966) *Pl. Physiol.*, **41**, Suppl. XV.
378a. SCHNEIDER, M. J. and STIMSON, W. R. (1971) *Pl. Physiol.*, **48**, 312.
379. SCHOPFER, P. (1966) *Planta*, **69**, 158.
380. SCHOPFER, P. (1967) *Planta*, **74**, 210.
381. SCHOPFER, P. (1972) In *Nucleic Acids and Proteins in Higher Plants*. Proc. Symp. Tihany (Hungary), ed. Farkas, G. L. *Symp. Biol. Hung.* **113**, 115.
382. SCHOPFER, P. and HOCK, B. (1971) *Planta*, **96**, 248.
383. SCHNARRENBERGER, C. and MOHR, H. (1970) *Planta*, **94**, 296.
384. SCHWABE, W. W. and WILSON, J. R. (1965) *Ann. Botany (London)* N.S., **29**, 383.
385. SETTY, F. and JAFFE, M. J. (1972) *Planta*, **108**, 127.

386. SIEGELMAN, H. W., CHAPMAN, D. J. and COLE, W. J. (1968) In *Porphyrins and Related Compounds*, ed. Goodwin, T. W. p. 107. Academic Press: London.
387. SIEGELMAN, H. W. and FIRER, E. H. (1964) *Biochemistry*, **3**, 418.
388. SIEGELMAN, H. W. and HENDRICKS, S. B. (1957) *Pl. Physiol.*, **32**, 393.
389. SIEGELMAN, H. W. and HENDRICKS, S. B. (1958) *Pl. Physiol.*, **33**, 185.
390. SIEGELMAN, H. W. and HENDRICKS, S. B. (1958) *Pl. Physiol.*, **33**, 409.
391. SIEGELMAN, H. W. and HENDRICKS, S. B. (1960) *Fed. Proc.*, **24**, 863.
392. SIEGELMAN, H. W., TURNER, B. C. and HENDRICKS, S. B. (1966) *Pl. Physiol.*, **41**, 1289.
393. SIEVERS, A. and VOLKMANN, D. (1972) *Planta*, **102**, 160.
394. SISLER, E. C. and KLEIN, W. H. (1961) *Physiol. Plant.*, **14**, 115.
395. SISLER, E. C. and KLEIN, W. H. (1963) *Physiol. Plant.*, **16**, 315.
396. SMILLIE, R. M. and SCOTT, N. S. (1969) *Progr. Mol. Subcell. Biol.*, **1**, 136.
397. SMITH, H. (1970) *Nature*, **227**, 665.
398. SMITH, H. (1970) *Phytochem.*, **9**, 965.
399. SMITH, H. (unpublished results)
400. SMITH, H. (1972) In *Phytochrome*, ed. Mitrakos, K. and Shropshire, W. Jnr. p. 433. Academic Press: London.
401. SMITH, H. (1973) In *Biosynthesis and its Regulation in Plants*, ed. Millborrow, B. V. p. 303. Academic Press: London.
402. SMITH, H. (1973) In *Seed Ecology*, ed. Heydecker, W. p. 219. Butterworths: London.
403. SMITH, H. and ATTRIDGE, T. H. (1970) *Phytochem.*, **9**, 487.
404. SMITH, H. and HARPER, D. B. (1970) *Phytochemistry*, **9**, 477.
405. SOROKIN, H. P. and THIMANN, K. V. (1960) *Nature*, **181**, 1038.
406. SPRUIT, C. J. P. (1970) *Meded. Landbouwhogeschool Wageningen*, **70**, 14.
407. SPRUIT, C. J. P. (1971) *Meded. Landbouwhogeschool Wageningen*, **71**, 21.
407a. SPRUIT, C. J. P. and KENDRICK, R. E. (1973) *Photochem. Photobiol.*, **18**, 145.
408. SPRUIT, C. J. P. and MANCINELLI, A. L. (1969) *Planta*, **88**, 303.
409. STAFFORD, H. A. (1948) *Am. J. Botany*, **35**, 706.
410. STEER, B. T. and GIBBS, M. (1969) *Pl. Physiol.*, **44**, 775.
410a. STEER, M. W., HOLDEN, J. H. W. and GUNNING, B. E. S. (1970) *Can. J. Genet. Cytol.*, **12**, 21.
411. STEINER, A. M. (1969) *Planta*, **86**, 334.
412. STEINER, A. M. (1969) *Planta*, **86**, 343.
413. STRAUS, J. (1959) *Pl. Physiol.*, **34**, 536.
413a. TAKATORI, S. and IMAHORI, K. (1971) *Phycologia*, **10**, 221.
414. TAKIMOTO, A. and HAMNER, K. C. (1965) *Pl. Physiol.*, **40**, 852.
415. TANADA, T. (1968) *Pl. Physiol.*, **43**, 2070.
416. TANADA, T. (1968) *Proc. Nat. Acad. Sci. U.S.*, **59**, 376.
417. TANADA, T. (1972) *Nature*, **236**, 460.
418. TANADA, T. (1972) *Pl. Physiol.*, **49**, 860.
419. TANADA, T. (1973) *Pl. Physiol.*, **51**, 150.
420. TANADA, T. (1973) *Pl. Physiol.*, **51**, 154.
421. TAYLOR, A. O. (1968) *Pl. Physiol.*, **43**, 767.
422. TAYLOR, A. O. and BONNER, B. A. (1967) *Pl. Physiol.*, **42**, 762.
423. TEZUKA, T. and YAMAMOTO, Y. (1969) *Bot. Mag., Tokyo*, **82**, 130.
424. TEZUKA, T. and YAMAMOTO, Y. (1972) *Pl. Physiol.*, **50**, 458.
425. THIMANN, K. V. and EDMUNDSON, Y. H. (1949) *Arch. Biochem.*, **22**, 33.
426. THOMAS, A. S. and DUNN, S. (1967) *Planta*, **72**, 198.
427. THOMAS, A. S. and DUNN, S. (1967) *Planta*, **72**, 208.
428. THOMSON, B. F. (1950) *Am. J. Botany*, **37**, 284.
429. THOMSON, B. F. (1951) *Am. J. Botany*, **38**, 635.
430. THOMSON, B. F. (1959) *Am. J. Botany*, **46**, 740.
431. THOMSON, B. F. and MILLER, P. M. (1962) *Am. J. Botany*, **49**, 383.
432. THOMSON, B. F. and MILLER, P. M. (1963) *Am. J. Botany*, **50**, 219.
433. THORPE, T. A., MAIER, V. P. and HASEGAWA, S. (1970) *Phytochemistry*, **10**, 711.
434. TOBIN, E. M. and BRIGGS, W. R. (1969) *Pl. Physiol.*, **44**, 148.
435. TOOLE, E. H., TOOLE, V. K., BORTHWICK, H. A. and HENDRICKS, S. B. (1955) *Pl. Physiol.*, **30**, 15.

436. TOOLE, V. K. and BORTHWICK, H. A. (1966) *Pl. Physiol.*, **41**, Suppl. xvii.
437. TOOLE, V. K., BORTHWICK, H. A., TOOLE, E. H. and SNOW, A. G. Jnr. (1968) *Pl. Physiol.*, **33**, S-23.
438. TOOLE, E. H., TOOLE, V. K., BORTHWICK, H. A. and HENDRICKS, S. B. (1953) *Pl. Physiol.*, **30**, 473.
439. TRAVIS, R. L., LIN, C. Y. and KEY, J. L. (1972) *Biochim. biophys. Acta*, **277**, 606.
440. TROYER, J. R. (1964) *Pl. Physiol.*, **39**, 907.
441. UNSER, G. and MASONER, M. (1972) *Naturwissenschaften.*, **59**, 39.
442. VALANNE, N. (1966) *Ann. Botan. Fennici*, **3**, 1.
443. VAN OVERBEEK, J. (1936) *Rec. Trav. bot. neerl.*, **33**, 333.
444. VAN POUCKE, M., CERFF, R., BARTLE, F. and MOHR, H. (1969) *Naturwissenschaften*, **56**, 417.
445. VINCE, D. (1964) *Biol. Rev.*, **39**, 506.
446. VIRGIN, H. J. (1961) *Physiol. Plant.*, **14**, 439.
447. VIRGIN, H. J. (1958) *Physiol. Plant.*, **11**, 347.
448. VIRGIN, H. J. (1962) *Physiol. Plant.*, **15**, 380.
449. VIRGIN, H. J., KAHN, A. and VON WETTSTEIN, D. (1963) *Photochem. Photobiol.*, **2**, 83.
449a. WAGNER, E., HAUPT, W. and LAUX, K. (1972) *Science*, **176**, 808.
450. WALKER, T. S. and BAILEY, J. L. (1968) *Biochem. J.* **107**, 603.
451. WAREING, P. F. (1966) *Botanical Society of British Isles Conf. Rep.* **No. 9**, p. 103, Pergamon Press: Oxford.
452. WEIDNER, M., JAKOBS, M. and MOHR, H. (1965) *Z. Naturforsch.*, **20b**, 689.
453. WEIDNER, M. and MOHR, H. (1967) *Planta*, **75**, 99.
454. WENT, F. W. (1928) *Rec. Trav. Botan. Neerl.*, **25**, 1.
455. WENT, F. W. (1941) *Am. J. Botany*, **28**, 83.
456. WENT, F. W. (1941) *Am. J. Botany*, **28**, 85.
457. WENT, F. W. (1957) *The Experimental Control of Plant Growth*, Chron. Bot.: Waltham, Mass.
458. WILKINS, M. B. (1965) *Pl. Physiol.*, **40**, 24.
459. WILKINS, M. B. and GOLDSMITH, M. H. M. (1964) *J. Exptl. Botany*, **15**, 600.
460. WILLIAMS, G. R. and NOVELLI, G. D. (1964) *Biochem. biophys. Res. Commun.*, **17**, 23.
461. WILLIAMS, G. R. and NOVELLI, G. D. (1968) *Biochim. biophys. Acta*, **155**, 183.
462. WILSON, J. R. and SCHWABE, W. W. (1964) *J. Exptl. Botany*, **15**, 368.
463. WITHROW, R. B. (1959) *Photoperiodism and Related Phenomena in Plants and Animals*, A.A.A.S., Washington.
464. WITHROW, R. B., KLEIN, W. H. and ELSTAD, V. (1957) *Pl. Physiol.*, **32**, 453.
465. WITHROW, R. B., KLEIN, W. H., PRICE, L. and ELSTAD, V. (1953) *Pl. Physiol.*, **28**, 1.
466. WITHROW, R. B. and WITHROW, A. P. (1956) In *Radiation Biology*, ed. Hollaender, A. Vol. 3, p. 126. McGraw-Hill: New York.
467. WITHROW, R. B., WOLFF, J. B. and PRICE, L. (1956) *Pl. Physiol.*, **31**, xiii–xiv.
468. YANIV, Z., MANCINELLI, A. L. and SMITH, P. (1967) *Pl. Physiol.*, **42**, 1479.
469. YOUNG, M. R., TOWERS, G. H. N. and NEISH, A. C. (1966) *Can. J. Bot.*, **44**, 341.
470. YUNGHANS, H. and JAFFE, M. J. (1970) *Physiol. Plant.*, **23**, 1004.
471. YUNGHANS, H. and JAFFE, M. J. (1972) *Pl. Physiol.*, **49**, 1.
472. KADMAN-ZAHAVI, A. (1960) *Bull. Res. Council Israel*, **9D**, 1–20.
473. ZIELKE, H. R. and FILNER, P. (1971) *J. Biol. Chem.*, **246**, 1772.
474. ZINSMEISTER, H. D. (1960) *Planta*, **55**, 647.
475. ZUCKER, M. (1963) *Pl. Physiol.*, **38**, 575.
476. ZUCKER, M. (1965) *Pl. Physiol.*, **40**, 779.
477. ZUCKER, M. (1968) *Pl. Physiol.*, **43**, 365.
478. ZUCKER, M. (1969) *Pl. Physiol.*, **44**, 912.
479. ZUCKER, M. (1970) *Biochim. biophys. Acta*, **208**, 331.
480. ZUCKER, M. (1971) *Pl. Physiol.*, **47**, 442.

Index